Transforming Food Waste into a Res

Transforming Food Waste into a Resource

Andrea Segrè and Silvia Gaiani
Faculty of Agriculture of the University of Bologna, Bologna, Italy

RSCPublishing

ISBN: 978-1-84973-253-6

A catalogue record for this book is available from the British Library

Published by The Royal Society of Chemistry,
Thomas Graham House, Science Park, Milton Road,
Cambridge CB4 0WF, UK

Registered Charity Number 207890

For further information see our web site at www.rsc.org

Printed in Great Britain by CPI Group (UK) Ltd, Croydon, CR0 4YY

Foreword: Wasting Food, Wasting People, Wasting the Earth

Industrial agriculture, industrial food processing, globalized trade and industrial retail have all been offered as interventions which increase efficiency and reduce food waste. The argument of waste reduction and increased food supply is the main justification for ecological agriculture being displaced by the Green Revolution and genetic engineering; artisanal food processing being displaced by industrial food processing; local and regional food systems being displaced by globalized, deregulated trade; and small and local retail being replaced by giant industrial retail. However, instead of reducing waste, industrialization and the corporate takeover of the food chain is creating waste at every level. It is wasting food, wasting people and wasting the Earth.

WASTING THE EARTH

Industrial agriculture is wasting the Earth both by failing to harvest its abundance and by eroding and polluting nature's capital—biodiversity, soil and water—thus undermining the Earth's capacity to produce food.

Industrial farming uses 10 kcal of energy to produce 1 kcal of food. This waste becomes entropy. Emissions of greenhouse gases (GHG) from industrial farming and globalized trade account for

Transforming Food Waste into a Resource
By Andrea Segrè and Silvia Gaiani
© Andrea Segrè and Silvia Gaiani, 2012
Published by the Royal Society of Chemistry, www.rsc.org

35–40% of all GHGs. Air pollution and atmospheric waste from industrial farming destabilize the climate, and climate instability affects food production and creates food insecurity. Industrial farming also uses 10 times more water than ecological agriculture and then pollutes water bodies with agricultural chemicals and waste from industrial factory farms. It has thus wasted the abundance of water.

Industrial farming has reduced the human diet from its use of 8500 plant species to 8 crops. In this way it has displaced biodiversity and thus reduced the potential nutrition available. It has wasted the abundance of biodiversity, destroyed the biodiversity in the soil and thus eroded soil fertility.

WASTING PEOPLE

Industrial food systems waste people through destroying livelihoods, and through wasting the knowledge and skills, evolved over millennia, held by farmers, women and indigenous peoples. Worse, food systems based on corporate profits rather than people's right to healthy, safe, nourishing and adequate food are wasting 1 billion people through hunger and another 2 billion through obesity and related diseases like diabetes and hypertension. Today every second child in India is 'stunted' and every fifth child is 'wasted' (low weight for height) even though India is an emerging economic superpower.

The 250 000 farmer suicides in India in a little over a decade are wasted lives. One million children dying annually for lack of food is a wasted future.

WASTING FOOD

Industrial processing and industrial retail waste food in multiple ways. Monocultures and uniformity lay waste healthy, nutritious food. Centralized and globalized distribution chains displace the diversity of foods in diverse local communities. The further food travels and the more it is industrially processed, the greater is the waste. As Jonathan Bloom has shown in *American Wasteland*, America throws away nearly half of its food. This food waste is totally unnecessary. Alternatives exists across the world that show ways to create food systems without a wasted earth, wasted people and wasted food.

CREATING ZERO WASTE FOOD SYSTEMS

Nature wastes nothing. Producing and distributing food in nature's ways is the most effective road to ending food waste.

- At the level of **production**, this means stopping the use of agricultural chemicals and promoting biodiverse ecological farming. It means biodiversity intensification and ecological intensification, instead of chemical intensification and pollution intensification.
- At the level of **processing**, this means promotion of artisanal processing instead of industrial processing.
- At the level of **distribution,** this means local markets and small, decentralized retail. Ecological, local, decentralized systems produce more food and no waste. They address the climate and biodiversity crisis, the hunger and food crisis, the unemployment and poverty crisis, simultaneously. Zero waste food systems honour the Earth and its people.

While we make this transition, we need to turn food waste into a resource. This is what Andrea Segré and Silvia Gaiani have offered in this book, *Transforming Food Waste Into A Resource.*

Vandana Shiva
Director of Research Foundation for Science,
Technology and Ecology,
Dehradun, India

Preface

Since civilization began, food and farming have told a story of our society, our values, and our relationship to the Earth.

Food is much more than simple nourishment: it is an expression of culture, culinary traditions, group identity and, at the same time, social status. Sharing food can mark entry into a community, making people an integral part of the same culture and putting them in communication with each other. A gift of food can create a bridge between people, and in all societies food has always played a role of major importance in social relations.

In the course of time food has profoundly changed and evolved: from being obtained from hunting and gathering, farming, ranching, and fishing, to being a mere commodity operated by multinational corporations using intensive farming and industrial agriculture methods with the aim to feed a growing world population. From being a source of energy, nutritional support and pleasure for the human body to being a controversial and debated terrain: if only fifty years ago food was simply 'food' now it has become 'junk food', 'slow food', 'fast food', 'whole food', 'organic food', 'genetically modified food'.

From the 1950s onwards, many traditional foods in the Western world have been removed from the shops shelves and replaced with thousands of new, aesthetically perfect foods and ready-made and pre-cooked meals. A real cornucopia of food, available all the time

Transforming Food Waste into a Resource
By Andrea Segrè and Silvia Gaiani
© Andrea Segrè and Silvia Gaiani, 2012
Published by the Royal Society of Chemistry, www.rsc.org

and in large quantities has been presented to consumers with varying tastes and differing purchasing power.

If on the one hand mass production has contributed to widespread food security, at least in developed countries, on the other hand it has skewed the traditional concept of demand, and generated a wildly dysfunctional market based on industrial agriculture. Never before in history has the food supply been so plentiful, so varied and cheap: the world has never produced so much food—in 2009 worldwide grain harvest was colossal, 5% above that of the previous year—and at the same time it has never consumed and wasted so much.[1]

In 2010, for the first time in history, the number of overweight people (around 1 billion) outnumbered that of the undernourished, hungry people (around 800 million).[2] The paradox is that the food produced globally could be enough to feed twice the world's population.

Paul Roberts, an American writer and long-time observer of energy issues and politics, depicts the global food market as a lumbering beast, organized on such a monolithic scale that it cannot adapt to the consequences of its own distortions.[3] Mass production of agricultural commodities, and too many years and subsidies spent in setting up big single-crop farms, have brought about efficiency but have also created a system full to bursting with overconsumption (and unequal distribution) on one hand and waste on the other.

The Western linear society, where extraction, production, consumption and waste are the normal, inevitable steps of a supposedly efficient mechanism, is no longer an option. In a world where food consumption accounts for one-third of all greenhouse gas emissions,[4] and where if everyone consumed as we do in the UK it would take six planets[5] to produce the amount of food needed, the present is unsustainable.

A new policy terrain should address old and fundamental questions. How much consumption is enough? Who needs to consume and waste less? Various scenarios are possible, from optimistic to pessimistic, strong to weak action, sustainable to crisis, costly now to unfundable in the long-term.[6]

Starting from an analysis of the contemporary food scenario, one that is marked by rising food prices, growing numbers of hungry people, an emerging politics of food scarcity, water shortage, declining resources, enormous quantity of wasted food and obesity,[7] our book aims at providing a framework of the new emerging trends in food policies.

As the title reveals, the focus of the book is specifically on food waste, a topic to which we have dedicated many years of research. Despite our long experience in food policy analysis and our hand-on experience of food waste, writing this book has been a fascinating experience and at the same time a real challenge. Precise global and national data on food waste are still lacking. It is difficult to compare the amount of food waste today with that in the past, because historical data have not been registered. Although the figures are notoriously poor and unreliable, they suggest that food waste takes place along the whole supply chain and that it is a world phenomenon, common to both developed and developing countries (although for different reasons), and compromising the overall food security.

One of the main reasons for losses is an increasing distance between the places where food is produced and those where it is consumed. Whereas in the past, many people produced their own food, now various parts of our diet come from food grown in many places in the world. In parallel with this trend, and closely associated with it, is the involvement of a growing number of actors and interests along the food chain. Farmers, the food processing industry, the marketing industry, transporters, shopkeepers, supermarkets, and consumers are all involved. Food waste is a market and policy failure—as well as a matter of education and awareness—that should not be neglected.

But food waste can also become a resource and an opportunity: re-use and recycle should turn into the new keywords of the 21st century.

The book is structured as follows:

Chapter 1 focuses on the evolution of food production and consumption, investigates if the world is entering a new food regime and presents the current global issues, *i.e.* food consumption, misconsumption, inequalities and the environmental impact of food production and consumption. Hypotheses concerning the future of food are then formulated and the concept of food waste, a neglected issue in current food policies, is introduced.

Chapter 2 attempts to define food waste and trace the historical reasons and cultural changes that have led to an increased amount of waste. Food waste is analysed along the food supply chain (at post-harvest stage, in the manufacturing and retail sector, at consumer level and at the level of businesses and institutions). Facts and figures are provided about food waste in Europe (UK and Italy in particular), Asia, the US and Africa. The environmental and socio-economic impacts of food waste are also analysed.

Chapter 3 presents a comparison between American and European food policies and food legislation, and analyses both governmental and non-governmental policies. The focus in on food waste laws. In this case the geographical areas considered are not only the US and Europe but also Japan, India and Brazil. The role of non-governmental organizationss and food activism in shaping the food scenario is also investigated.

Chapter 4 is dedicated to the presentation of food waste prevention strategies and the analysis of food disposal solutions. The prevention of food waste, in order to be effective, requires a multilevel approach where agricultural, national and international appropriate policies and strategies integrate with consumer awareness and effective technologies.

The authors illustrate the most widespread food waste disposal options (dumping, burning and minimizing) and investigate whether forcing retailers to declare their annual waste quantities and reveal their waste quantification methodologies could be a good way to help the reduction of food waste.

Examples of technologies that, in both developed and developing countries, can contribute to improve shelf life for perishable foods and semi-prepared meals, transform food waste into energy or simply track food waste are then presented. The chapter concludes with speculations concerning a possible future scenario and policy recommendations that could be adopted to reduce food waste.

Chapter 5 presents food waste recovery programmes considered by the authors as a viable alternative to landfilling and a way to help people in need. Selected case studies from the US (school food service programmes), UK (FareShare, a charity relieving food poverty), Denmark (Stop Wasting Food, a Danish consumer movement) and France (A.N.D.E.S— the Association Nationale de Dévelopement des Epiceries Solidaires) are provided as evidence. The benefits and challenges of such programmes are discussed and various international approaches to food reduction are presented.

Chapter 6 focuses on food waste in Italy. Last Minute Market, a successful project launched in 2000 at the Faculty of Agriculture of the University of Bologna by Professor Andrea Segrè, one of the authors of the book, is presented. Last Minute Market is considered to be a 360° action against waste and a win–win model as all its stakeholders obtain benefits. The operation of Last Minute Market is explained in detail and its nutritional, economic and environmental impacts are presented. We conclude the chapter by presenting the Joint Declaration against Food Waste and the project 'A Year Against Food Waste', two European initiatives launched by the authors and supported by Last Minute Market staff whose aim is to raise awareness about food waste and involve the European Commission, as well as international institutions like the United Nations, to take position and develop common action to reduce by 50% the global food waste by 2025.

The book concludes by stressing the need for a new comprehensive, integrated, multisectoral and multilevel action that focuses on sufficiency, transparency and efficiency to reduce food waste. The world is entering a time where big principles will come into play. Staying on familiar territory is no longer an option. Governments need to be more proactive, the supply chain needs better goal-setting, food waste and overproduction must be reduced, consumers must be well informed.

A shift from quantity to quality, from induced needs to real necessities, should soon take place. The readers are invited to develop an ecological intelligence, defined as the ability to learn from experiences and interact effectively with the environment by reducing in an effective way the environmental impacts on the resources.

REFERENCES

1. IFPRI, *Climate Change Impact on Agriculture and Costs of Adaptation Report*, Washington, 2009, p. VII available at http://www.ifpri.org/sites/default/files/publications/pr21.pdf
2. C. Nordqvist, *Overweight People Now Outnumber Hungry People*, Medical News Today, 2010.
3. P. Roberts, *The End of Food*, Bloomsbury, 2008, p. 3.
4. A. Tukker, *Environmental Impact of Products (EIPRO): Analysis of the Life Cycle Environmental Impacts Related to the Final*

Consumption of the EU-25, EUR 22284 EN. 2006, European Commission Joint Research Centre, Brussels.

5. E. Audsley, M. Brander, J. Chatterton, D. Murphy-Bokern, C. Webster and A. Williams, *How Low Can We Go? An Assessment of Greenhouse Gas Emissions from the UK Food System and the Scope to Reduce them by 2050*, FCRN-WWF-UK, 2009.

6. B. R. J. Ravetz and T. Wiedmann, *Footprint North West: A Preliminary Ecological Footprint of the North West Region*, Stockholm Environment Institute (York), Centre for Urban and Regional Ecology, Action for Sustainability (North West Regional Assembly) Manchester and York, 2005.

7. M. Nestle, *Food Politics: How the Food Industry Influences Nutrition and Health*, University of California Press, (revised edition 2007).

Acknowledgements

This book would not have been possible without the support of many people.

Deepest gratitude is due to the Last Minute Market staff, Luca Falasconi, Matteo Guidi, Silvia Marra, Anastasia Scotto, Eleonora Morganti and Alessandro Politano, who supported us with empirical data and without whom Last Minute Market would not exist.

We wish to express our gratitude to Paolo De Castro, Chair of the European Parliament's Committee on Agriculture and Rural Development, former two-term Italian Agriculture Minister and Professor of Agricultural Economics, for his valuable support in making the conference 'Transforming Food Waste into a Resource' at the European Parliament in 2010 possible.

We would also like to convey thanks to Professor Jan Lundqvist, Professor and Senior Scientific Advisor, Stockholm International Water Institute (Sweden); Professor Paul Connett, Professor Emeritus in Environmental Chemistry, St Lawrence University, New York (USA) and Director of the American Environmental Health Studies Project (AEHSP); and Professor Tim Lang, Professor of Food Policy at the City University of London, UK, for the sharing of the literature and for their encouragement.

Special thanks also to Tony Lowe and Maria Ohlson from FareShare (UK), Selina Juul from the Stop Wasting Food movement (Denmark), Tory Coates from FoodCycle (UK), Kate Bull

Transforming Food Waste into a Resource
By Andrea Segré and Silvia Gaiani
© Andrea Segré and Silvia Gaiani, 2012
Published by the Royal Society of Chemistry, www.rsc.org

from The People's Supermarket (UK), Agathe Cousin and Guillaume Bapts from A.N.D.E.S (France) and Emily Teel, a master student from the University of Gastronomic Sciences in Colorno (Italy).

We also wish to express gratitude to our publisher, RSC, and in particular to Sue Humphreys and Alice Toby-Brant, to the illustrator of the book cover, Francesco Tullio Altan, and last but not least to our beloved families for their understanding and support while we were writing this book.

Andrea Segrè
Silvia Gaiani

Contents

Transforming Food Waste into a Resource
By Andrea Segrè and Silvia Gaiani
© Andrea Segrè and Silvia Gaiani, 2012
Published by the Royal Society of Chemistry, www.rsc.org

CHAPTER 1

The R(evolution) in Global Food Production and Consumption

One North American consumes about the same amount as 30 Bangladeshis. The average American consumes about 3747 kcal per day (not including junk food) compared to the recommended 2000 to 2500 kcal per day. Globally, while 0.9 billion suffer from malnutrition by eating too little, there are about 1.6 billion who suffer from eating too much.

(http://peoplesfoodpolicy.ca/chapter6D11)

1.1 THE EVOLUTION OF FOOD AND FOOD PRODUCTION

By most accounts the story of food started about 3 million years ago, with the first appearance of humans on the planet,[1] and changed as the human race evolved, climate changed and technological progress brought about new inventions and tools.

Originally the human diet was mainly herbaceous, based on fibrous plant matter, fruits and vegetables: it began to change only 500 000 years ago when the large, more upright *Homo erectus*, started to use weapons to hunt rodents, reptiles and even small deer. A sudden change in climate, which caused a decrease in temperatures, led human beings to change their diet and search for

Transforming Food Waste into a Resource
By Andrea Segrè and Silvia Gaiani
© Andrea Segrè and Silvia Gaiani, 2012
Published by the Royal Society of Chemistry, www.rsc.org

more calorie-rich, highly nutritious food. It was at that time that animal food started to become essential in the human diet: muscle, fat, brains and organs made up 65% of the total calories humans ate at that time.[2] Moreover, animal foods were easier to digest— animal and human tissues have the same 16 amino acids, animal converts easily into human—and provided extra energy that could be used for hunting.

About 90 000 years ago, by the start of the last Ice Age, hunting was the main occupation. Research suggest that humans at that time needed hours and hours of effort, over many miles of territory, to find enough to eat. Daily life was at that time very brutal and solitary, with low birth rates and low average life expectancy.[3] The population was around 1 million and did not grow for tens of thousands of years.

As the centuries went by, it became clear that the 'hunter-gatherer' mode[4] was insufficient and that a new food strategy was required; at some point our ancestors began to produce food, thus shifting from hunting to farming. The shift was accompanied by a deep knowledge of plants developed over the years by our ancestors, and by the beginning of social organization necessary to tackle the complex and large-scale task of farming.

The discovery of fire marked a major step in the ability of humans to manipulate nature. Used for many purposes—heat, light, protection from wild beasts, sending messages and drying clothing—fire made possible progressive cultural developments that were enormously important, especially in terms of diet. For Levi Strauss,[5] cooking food using fire is the invention that made human beings human. Before we learned about cooking, food (and especially meat) was eaten raw, spoiled or rotten. The use of fire brought about a decisive change. In Levi Strauss's structuralist approach, cooking symbolically marks a transition from nature to culture, and also from nature to society, given that while 'raw' is natural in origin, 'cooked' implies a step that is both cultural and social.

Language may also have been developed to reduce tension connected with the division of food. At the origin of what we today call conviviality was the primitive practice of sharing food around the fire by groups of human beings who sat face-to-face, smiling, laughing—and, with time, conversing. These practices were not found among other species, not only because of their fear of fire, but because within the animal kingdom, direct eye contact, opening the mouth and showing teeth are typically hostile gestures.

Further changes were brought about by the first agricultural revolution which began in different places—Central Asia, America and southeast Asia—and at different times, but the circumstances that brought to its development were probably very similar: between 10 000 BC and 6000 BC small groups started to grow wheat in the Middle East, corn in Mesoamerica and rice in Asia.[6] It is supposed that by 5000 BC agriculture had reached every continent except Australia.[7]

The Fertile Crescent of the Middle East in particular was home to a variety of edible and easily cultivated crops—wheat and barley among the cereal crops, and lentils, peas and chickpeas among the vegetables—and was endowed with wild goats, sheep, pigs and cattle, all of which were domesticated and became important sources of both food and fertilizers. Having the most favourable combination of plants and animals eventually translated into a significant cultural advantage for the Middle East and Europe.

The more concentrated form of food production required by agriculture—farmers could in fact produce enough food to feed large groups using a small area of land compared to hunter-gathers who needed square miles to collect enough food for survival—allowed the growth first of villages and then of cities.

By 3500 BC Egyptian farmers were producing more grain than they could eat themselves,[8] thus accumulating surpluses. Surpluses provided food security and at the same time allowed the accumulation of wealth and an impetus toward development: urbanization, a high degree of economic specialization and social inequality.[9] Food surpluses allowed people to specialize in tasks different from agriculture and making food.

The city of Rome was highly dependent on wheat from Egypt and North Africa to supply the grain that was distributed free of charge to its plebeians. The annona (the distribution of free or reduced-price grain or bread) had reached impressive dimensions by 350 BC: an estimated 120 000 people received 6 half-pound loaves per day, provided by 274 public bakeries. It was one of the world's first examples of mass production of a specific food product.

In the evolving food economy, food security became a question of power and political influence, of domination and military control—exactly as it is at the present time. When the Roman

Empire collapsed in the 4th century AD, the Western food economy collapsed so completely that for the next six centuries the global population rose from 300 million to just 310 million.[10]

Things changed in the agrarian system when a series of extraordinary innovations were brought about at the beginning of the 8th century: a new plough capable of breaking the dense, wet soils of northern Europe reached Germany and opened up a major new grain source for the rest of the continent. Grist mills powered by wind or water appeared up all over Europe, thus providing large-scale processing of grain into flour.[11] At the same time biology and chemistry began to take shape thus allowing a better understanding of the natural systems and the development of better agricultural techniques.

In the later Middle Ages, the revitalizing power of legumes, which supply nitrogen to the soil, a technique lost since Roman times, was rediscovered. Rotating fields through grain, legumes and fallow boosted productivity by at least one-third and added peas, beans, chickpeas, lentils and other vegetables to the European diet.

China, often thought of as a land of rice, also depended heavily on millet, wheat and soybeans. Rice production increased significantly in the 11th century when new strains were imported from southeast Asia. Chinese fishermen also gathered fish from the ocean, lakes and rivers, and sold them in vast central markets, which supplied networks of cookshops, restaurants, banqueting halls and other eating places.[12]

The Arab world as well had a varied and sophisticated system of food production, with water-powered mills grinding grain full-time in North Africa and fishermen packing Mediterranean tuna in salt. The Arabs introduced citrus, rice and sugarcane to Europe and controlled the spice trade with India.[13] European interest in breaking the Arab hold on the spice trade led to the voyages of discovery of Vasco da Gama and Columbus.

In the early 13th century, the first food regulatory law, the Assize of Bread, was proclaimed by the king of England:[14] it prohibited bakers from mixing ground peas and beans into bread dough, thus starting the game between the food industry and the consumers.

The discovery of the New World touched off the greatest and most rapid spread of new crops the world had seen. The Americas contributed maize, potatoes, tomatoes and peppers to Europe, while the Europeans brought wheat and other staple crops to

Brazil and later the Caribbean region. Sugarcane cultivation created a demand for labour that was met by the African slave trade. The 'Columbian Exchange'[15] thus laid the basis for much of the subsequent economic and political history of the New World.

In Europe, the decline of feudalism and the rise of cities and towns helped move agriculture from subsistence to a market orientation. Land that had been held in common and used mainly for grazing was consolidated under the control of individual landowners, which greatly increased production of both crops and animals. The draining of marshy land, especially in England and the Low Countries, was accelerated. All these trends supported the more intensive cultivation of the available land and the production of more and cheaper food for growing and more urban populations.

By 1700, European agriculture could provide approximately two-and-a-half times the yield per input of seed that had been normal in the Middle Ages.[16] As a consequence, world population jumped from roughly 500 million to 800 million. Italy and France had population densities greater than their farmlands could sustain, and China and India were not behind.[17] The result of the demographic tension brought about famines and local shortages, and people migrated in search of food.

In 1798 Thomas Malthus, an Anglican clergyman and economist, explained his theory in a treatise entitled *An Essay on the Principle of Population*.[18] Malthus believed that hunger would never be eliminated because any increase in food would serve only to make the population bigger. Since crop yields can increase only linearly whereas population increase geometrically (doubling every several hundred years), Malthus reasoned that population growth would soon outpace humankind's capacity to feed itself. His calculations suggested that a cataclysmic famine would have happened by the middle of the 19th century. Fortunately, science and technology, which played an increasingly important role in food production in the 18th and 19th centuries, proved Malthus's forecast to be wrong.

The mechanization of agriculture advanced rapidly in the 19th century, with mechanical reapers, the tractor, and electric milking machines, among other innovations. Scientists developed a better understanding of the nutritional components of food, which led to

an emphasis on a balanced diet and, by the 20th century, resulted
in the improvement of food with the addition of vitamins and
minerals to products such as bread and breakfast food.[19]

It was finally with the emergence of an international food system,
built on railways, shipping routes, new preservation technologies
and above all free trade, that the growing food demand from
Europe could be linked with distant suppliers in North and South
America and the prototype of the modern food system began to
develop. Along with good infrastructure, preservation is the other
aspect that marked a change in the food supply chain. The preservation of food by heating it and sealing it in jars or cans began in
the early 19th century, followed by pasteurization of wine and later
milk to kill spoilage organisms. Canning and pasteurization made a
wider variety of foods available to urban populations. In 1900
Britain was importing almost half of its calories. For centuries
nations had sought to produce as much of their food as possible,
importing food only in hard times, but with the advent of free trade
and improved transportation systems, the idea of self-sufficiency
became obsolete and impractical.

With the development of steamships and refrigeration in the 19th
century, worldwide food exports went from 4 Mt in the 1850s to
18 Mt 30 years later and 40 Mt by 1914.[20] In 1885 more than half of
the population of the United States was engaged in farming[21] and
the average American family spent half of its household income
on food.

In 1900 some families sat down to a plate of potatoes for their
main meal, and malnutrition was common among poor children.
Food was also expensive. In 1914 a working class family spent
about 60% of its income on food. By 1937 food was cheaper and
families only spent about 35% of their income on food.[22] Moreover sweets, which were a luxury in 1914, became much more
common in the 1920s and 1930s. The diet of ordinary people
improved greatly during the 20th century as agriculture, fisheries,
livestock and poultry production progressed and became very
efficient in Europe, North America, Australia, Argentina, Brazil
and Japan (but lagged in regions such as Africa, where antiquated
farm methods barely kept pace with rising populations).

During the Second World War the demand for safe, transpor-
table, easy-to-eat military rations spurred the food industry to
come up with innovations in formulation, preservation techniques

and packaging. Food processing became more competitive and demand for convenience food more global. Also, during the war and immediately after it much food—such as butter, sugar and meat—was rationed. After these restrictions were lifted, convenience foods became common in the late 20th century and the Western diet also became more varied. Several new foods, such as hot dogs, sliced bread, instant coffee, fish fingers and fruit-flavoured yoghurt, were invented.[23]

Also, the way people shopped changed. In the early 20th century people usually went to small local shops such as a baker or butcher. In the 1950s and 1960s supermarkets replaced many small shops. During this time a dietary model described as 'American', very heavy in beef, came to be part of the mode of consumption, thus creating an incentive for industrial livestock production. Food became a durable consumer good and new, frozen foods were invented alongside meals such as TV dinners.

From the point of view of consumption, agriculture changed in two important ways: maize production increased as grain production shifted from human to animal feed, and many farms became suppliers of raw material for industrial food manufacturing. In the early 1960s, most nations were self-sufficient in food. In the period 1950–1984, the introduction of high-yield crops and energy intensive agriculture ushered in the Green Revolution, leading to further increased crop production—in some cases by 100-fold or more—and, except in parts of Africa, production exceeded population growth throughout the world.[24]

Plant breeding was principally aimed at designing plants that could tolerate high levels of fertilizer use and improving the harvest index. But at the same time the Green Revolution was technologically suited to special circumstances and implemented in a manner that has not proved to be environmentally sustainable. The technology enhanced soil erosion and polluted groundwater and surface-water resources, and the increased use of pesticides caused serious public health and environmental problems.

As people moved into the cities in the 1960s, they lost much of their food independence. Families became reliant on the food industry to provide them with what they needed. Even farmers moved away from growing their own food and home food production as they began to focus on growing cash crops. By the middle of the 20th century farming was transforming in ways it never had. Private family farms were quickly diminishing and being replaced with large, corporate, monocultural farms.

While demand for commercially processed foods increased and accessibility to food in general also increased, obesity and other health problems began to be an issue despite the transformation of meals from the 'meat and potato' standard to lighter fare in the mid-1920s and 1930s.[25]

As incomes increased and food, homes and technology became more affordable in the 1950s, people continued to consume greater and greater amounts of commercially processed foods and ready-made meals.[26] By 1975 the average amount of time spent in meal preparation was only 10 hours compared to 44 hours or more in 1900. Through the 1980s the trend for convenience food continued. It was easy to purchase every meal out with the accommodating fast-food restaurants and for those who still chose to eat at home, the grocery store was packed full of pre-cooked and easy-to-prepare dishes. The evolution in food production, distribution and consumption has never changed so fast as in the last 50 years.

If food production has increased tremendously, it is also true that much of what is now eaten in the West is not food so much as, in Michael Pollan's[27] terms, stuff that is merely 'foodish', food of an inferior quality, and that much of the apparent abundance of choice available to the affluent Western consumer is an illusion. According to Pollan, most of what we eat is derived from the high-yield, low-maintenance crops that the food industry prefers to grow, and sells to us in myriad 'foodish' forms. For example, 75% of the vegetable oils in our diet come from soy (representing 20% of daily calories) and more than half of the sweeteners we consume come from corn (representing around 10% of daily calories).

1.2 IS THE WORLD ENTERING A FOURTH FOOD REGIME?

The 20th century is certainly a time that has brought about unprecedented and rapid change in the food system. Harriet Friedman, professor of sociology at the University of Toronto, suggests dividing the century into three main periods or food regimes.[28] A food system is a chain of activities from production ('the field') to consumption ('the table'), with particular emphasis on processing and marketing and the multiple transformations of food that these entail.[29]

The **first food regime** (pre-1914), also called the 'colonial-settler diasporic food regime', was based on extensive forms of capitalist production relations under which agricultural exports from white settlers in countries in Africa, South America and Asia supplied unprocessed and semi-processed foods and materials to metropolitan states in America and Europe. Trade was multilateral and the world economy was based on an international community.

The **second food regime** (1947–1970s), also called the 'mercantilist food regime', was a period characterized by intensive forms of capitalist production relations involving the modernization and industrialization of farming. Rising standards of diet were accompanied by food wars and food aid.

The **third food regime** (1970s onwards), also called the 'corporate food regime', is based upon the so-called 'clean and green' aspects of food—that is, freshness and 'naturalness'—while at the same time promoting a 'corporate-environmental food regime', and the emergence of so-called 'green capitalism'.

The third food regime—which according to Friedman should be the contemporary regime—is supply driven just like the first and second food regimes: food production, rather than consumption, lies at its core, but it differs from previous food regimes in the role of global capital and in its organizational form.

The modern food system is governed by a neoliberal mode of regulation, characterized by flexible production and the international sourcing of a wide and diverse range of food products on terms set by the international retailers, and increasingly organized around a set of concerns based on convenience, choice, health and 'wellness', freshness and innovation.

Food is increasingly being consumed in the North as a consequence of the cultural construction of capitalist consumer society: most of the time consumers are not able to understand where the products they purchase come from; they consume the food because of advertising, which allows capital to realize the surplus value that is created in the production process.

The general global trends in modern food systems are well documented and are summarized in Table 1.1.[30] The table illustrates the multiple actors involved in food systems, the broad array

Table 1.1 Comparing traditional and modern food systems

Food system feature	Traditional food systems	Modern food systems
Principal employment in food sector	In food production	In food processing, packaging and retail
Supply chain	Short, local	Long with many food miles
Food production system	Diverse, varied productivity	Few crops predominate
Typical farm	Family based	Industrial, large
Typical food consumed	Basic staples	Processed food with a brand name
Purchased food bought from	Small local shops or markets	Large supermarket chain
Nutritional concern	Undernutrition	Chronic dietary diseases
Main source of national food shocks	Poor rains, production shocks	International prices and trade problems
Main source of household food shocks	Poor rains, production shocks	Income shocks leading to food poverty
Major environmental concerns	Soil degradation	Nutrient loading, water demand, greenhouse gas emissions
Influential scale	Local to national	National to global

Source: adapted from Maxwell and Slater (2003).

of environment and social interactions they encompass and the multiple policy challenges posed.

The question now is whether we have entered a **fourth food regime**, characterized on one hand by environmental concerns over water availability, pollution from agricultural inputs and soil loss[31] and a large increase in the energy demands throughout the food production sectors,[32] and on the other by deep concerns for health, the revival of handcrafted or specialty food, regional dishes, niche markets, the rise of biotechnology and nanotechnology and a shift towards private power at the expense of public responsibility for managing surpluses.

It is evident that farming is no longer the dominant economic activity in the overall food system. Among the main areas of activities are now distribution and retail: food travels very long distances and the role and number of supermarkets is rapidly increasing, with considerable vertical and horizontal concentration among the major owners,[33] a trend for the retail sector as a whole.

Overall growth in incomes has caused a worldwide dietary transition to more meat (with a concomitant rise in demand for grain production), dairy, sugars and oils. Consequently, nutrition concerns relate to malnutrition in some places and obesity in others, as there is inequitable distribution of the quality as well as quantity of food, and negative consequences arise from multiple eating patterns.[34] This is exacerbated by the growth of urban populations who rely almost completely on purchasing food[35] and waste enormous amount of products that are still edible. It is an highly imbalanced, unfair and unsustainable food system which requires better and greener policies, ecological intelligence, parallel models of development alongside increased efforts in reconstructing choice, community and social relation to food. Food is now primarily an issue of socio-political arrangements and norms that govern access to the resources needed for food production and markets, not technical ability to produce more food.

1.3 THE CURRENT GLOBAL ISSUES—FOOD CONSUMPTION, MISCONSUMPTION AND INEQUALITIES

As stated by Jan Lundqvist in his policy brief *Saving Water: From Field to Fork—Curbing Losses and Wastage in the Food Chain*,[36] what we now face is indeed a time of contradictory tendencies. In 2008 the World Bank argued in its World Development Report that by 2030 cereal production needs to increase by nearly 50% and meat production by 85% to meet the expected growth in demand. Conclusions at the World Food Summit, in November 2009, stated the need to increase food production by 70% by 2050, envisaging a striking acceleration in the population boom and the commitment to reflect a shift from aid to the development of agriculture in poor countries.

According to latest official population estimates, in 2009 the world population was calculated at 6 790 062 216. By 2050 it is expected to reach 9.4 billion.[37] However, regardless of alarmist publications such as Lester Brown's *Could Food Shortages Bring Down Civilization?*,[38] where the author argues that a new food shortage could eliminate our civilization exactly as it happened to the Sumerians, Mayans, and many others, the threat of insufficient food for a growing population is probably a false alarm. In line with recent studies,[39] it seems that the current food production would be adequate to feed 1.5 times the world population, although with a mainly vegetarian diet. What is consequently most striking is that, in a time of global significant improvements in raising food consumption per person, both food production and undernourishment are increasing exponentially.

Michael Pollan's 'omnivore's dilemma'—the dilemma arising from having the choice to eat just about anything nature has to offer, thus leading inevitably to anxiety—is not common to all the regions of the world. Between 2007 and 2008 there was a dramatic jump in the number of undernourished people: an estimated 850 million people are now lacking adequate nourishment and access to food.[40] Some developing countries (especially in sub-Saharan Africa, *e.g.* Somalia, Burundi, Rwanda and Kenya) have declined further from what was already a very low *per capita* food consumption level. Perhaps surprisingly, food insecurity is most prevalent among rural populations[41] and women are still the poorest of the world's poor, representing 70% of the people who live in absolute poverty.

It is relevant to underline that production and supply are very different from access and beneficial consumption. There is a striking correlation between areas with a high proportion of undernourished people and a high proportion of the population who are extremely poor: poverty, conflicts, inappropriate food policies and adverse weather conditions are the main causes that prevent hundreds of millions of people from being food secure.[42] The recent rising price of food is another burning issue: families in developing countries spend up to 70% of their income on food, compared with 5–10% in the US.[43]

A joint report from the Organisation for Economic Co-operation and Development and the Food and Agriculture Organization (FAO) of the United Nations[44] projects that food commodity prices will continue to rise—about 10–20% in real terms in the next decade—mainly because farming is underinvested, and the infrastructure connecting producers with global markets is inadequate. The increases in agricultural commodity prices, in particular the prices of wheat, coarse grains, rice and oilseed crops, have been a significant factor driving up the cost of food.

What is particularly impressive is the imbalance between the richest and the poorest people in the world. The richest one-fifth of the world consume 45% of all meat and fish, the poorest a mere 5%. The richest consume 58% of total energy, the poorest less than 4%. The richest consume 84% of all paper, the poorest 1.1%.[45] Russia now has the world's greatest inequality, with the richest 20% having 11 times the income of the bottom 20%. Income inequalities have also grown dramatically in China, Indonesia, Thailand, other East and Southeast Asian countries, and in the industrialized countries, especially Sweden, Britain, and the US.[46] Using the latest figures available (2005), the wealthiest 20% of the world accounted for 76.6% of total private consumption, the poorest fifth just 1.5% (Figure 1.1).[47] Breaking that down slightly further, the poorest 10% accounted for just 0.5% and the wealthiest 10% accounted for 59% of all the consumption (Figure 1.2).

Today, the consumption–poverty–inequality nexus is accelerating. If the trends continue without change—not redistributing from high-income to low-income consumers, not shifting from polluting to cleaner goods and production technologies, not promoting goods that empower poor producers, not shifting priority from consumption for conspicuous display to meeting basic needs—today's problems of consumption and human development will worsen.[48] The real issue is not consumption itself but its patterns and effects.

It is worth reporting here some global expenditures on a range of several products that reflect world priorities (Table 1.2).

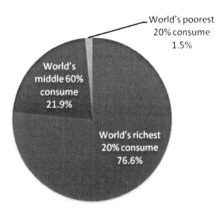

Figure 1.1 Share of world's private consumption, 2005. Source: World Development Bank Indicators, 2008.

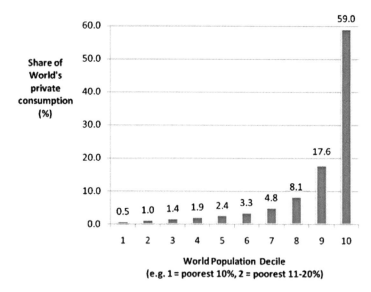

Figure 1.2 Inequality of consumption, 2005. Source: World Development Bank Indicators, 2008.

Table 1.2 Global priorities

Global priority	*$US billions*
Ice cream in Europe	11
Perfumes in Europe and the United States	12
Pet foods in Europe and the United States	17
Business entertainment in Japan	35
Cigarettes in Europe	50
Alcoholic drinks in Europe	105
Military spending in the world	780
Basic education for all	6
Water and sanitation for all	9
Reproductive health for all women	12
Basic health and nutrition	13

Source: The State of Human Development, United Nations Human Development Report 1998, Chapter 1, p. 37.

'Overpopulation' is usually blamed as the major cause of environmental degradation, but the statistics strongly suggests otherwise. As we will see, consumption patterns today do not meet everyone's needs. The system that drives these consumption patterns contributes to inequality of consumption patterns too.

1.3.1   Inequality in the *Per Capita* Food Supply

Socio-economic analysis throughout the world demonstrates that, although there might be sufficient food supplies available in a society, access can vary significantly between groups of people and also within a household. Even with a public distribution system in place and food available in stores, there may be people who are food insecure. An analysis of food supply data and the incidence of undernourishment in the world reveals a direct and linear reduction in the number of undernourished people with increased food supply.

The most comprehensive database for such calculations is FAO's Food Balance Sheets[49] (FBS), which provides information for individual countries on production, net exports or imports and non-food use of food. The quality of data depends on reports from the individual country. These sets of data can be used to estimate the supply of food on a country basis. They do not, however, show how much food is lost, wasted or eaten, and almost invariably result in an overestimation in food consumption compared with dietary surveys at the individual level.[50]

The FBS data do not provide information on the variability within areas of a country or between different socio-demographic subgroups in the population, as those data are provided by individual cross-sectional dietary surveys at the national level. Furthermore, FBSs do not give any indication of the differences that may exist in the diet consumed by different population groups, *e.g.* different socio-economic groups, ecological zones and geographical areas within a country, nor do they provide information on seasonal variations in the total food supply. To obtain a complete picture, food consumption surveys showing the distribution of the national food supply at various times of the year among different groups of the population should be conducted. Nonetheless, only the FBS data can show long-term trends in food availability for a large number of countries as they are available for every country in the world, for every food item.

Attempts have been made to estimate the minimum dietary energy requirement. According to FAO,[51] for instance, these estimates vary from 1730 to about 2000 kcal *per capita* per day for various countries. According to the Swaminathan Research Foundation,[52] the average food intake that is 70% of the international norm for food security, *i.e.* $0.7 \times 2700 = 1890$ kcal per day, may be an acceptable figure. What is generally acceptable must be related to nutritional and medical criteria. It is also related to the age and occupational structure of the population, among other things. Smil[53] provides examples showing that food intake at *per capita* levels below 2000 kcal per day has not resulted in documented signs of undernourishment. A more appropriate term for this variable would be 'national average apparent food consumption' since the data come from the national FBS rather than from food consumption surveys.

Analysis of FAOSTAT data[54] shows that dietary energy measured in kcal *per capita* per day has been steadily increasing on a worldwide basis; availability of calories *per capita* from the mid-1960s to the late 1990s increased by approximately 450 kcal per day globally, and by over 600 kcal per day in developing countries. This change has not, however, been equal across regions. The *per capita* supply of calories has remained almost stagnant in sub-Saharan Africa and has recently fallen in the countries in economic transition. In contrast, the *per capita* supply of energy has risen dramatically in east Asia (by almost 1000 kcal *per capita* per day, mainly in China) and in the Near East/North Africa region (by over 700 kcal *per capita* per day).

Some countries, like the US, conduct national surveys on a regular basis as a government requirement, but most do so less frequently because of the large costs involved. In spite of this, national surveys are the major source of reliable information on actual dietary intake around the world. These are often supplemented by small surveys in single locations on smaller numbers of individuals.

Difficulties exist in making international comparisons in food intake as a result of variations in the methodology involved in ascertaining food intake. Many countries, particularly in the developing world, do not have the resources to mount individual-level nutrition surveys. Such surveys are prohibitively expensive and labour-intensive. Also, differences in methodology and data analysis greatly curtail our ability to make meaningful comparisons of results at the international level.

Although the *per capita* food consumption projections foresee a rise of almost 100 kcal per day—going from almost 3000 kcal per day per person per in 2010 to 3130 in 2050 (Table 1.3)[55]—dramatic improvements and better food policies are nevertheless required.

Table 1.3 *Per capita* food consumption (kcal per person per day)

	1969–1979	1979–1988	1989–1991	1999–2001	2015	2030	2050
World	2411	2549	2704	2950	2789	3040	3130
Developing countries	2111	2308	2520	2654	2860	2960	3070
Sub-Saharan Africa	2100	2078	2106	2194	2420	2600	2830
Near East/ North Africa	2382	2834	3011	2974	3080	3130	3190
Latin America/ Caribbean	2465	2698	2689	2836	2990	3120	3200
South Asia	2066	2084	2329	2392	2660	2790	2980
East Asia	2012	2317	2625	2872	3110	3190	3230
Industrial countries	3046	3133	3292	3446	3480	3520	3540
Transition countries	3323	3389	3280	2900	3030	3150	3270

Source: Alexandratos (2006).

Despite the fact that the aggregate value of world agriculture, including all food and non-food crops and livestock commodities, has been growing by 2.1–2.3% annually on average during the last four decades in terms of calories arising from different major food commodities, large differences can still be seen between the developing and industrial countries. The marked rise in available food energy observed globally has been accompanied by changes in the composition of the diet, and more changes are about to come in the next few years. Although global figures and trends could be deceptive, it is noteworthy that the process involved in dietary changes over the years appears to follow a pattern involving two main stages.

In the first stage, known as the '**expansion**' effect, the main change is in terms of increased energy supplies, with these extra calories coming from cheaper foodstuffs of vegetable origin.[56] This development has been ubiquitous, occurring in both developed and developing countries. The second stage, called the '**substitution**' effect, results in a shift in the consumption of foodstuffs with no major change in the overall energy supply. This shift is primarily from carbohydrate-rich staples (cereals, roots, tubers) to vegetable oils, animal products (meat and dairy foods) and sugar. In contrast to the first stage, this one is country-specific and is influenced by culture, beliefs and religious traditions. In particular, such traditions can influence the extent to which animal products substitute vegetable products and the specific types of meat and animal products consumed. For example, meat demand is growing faster in China than in mostly vegetarian India. The *per capita* supply of meat in India seems to remain relatively low, projected at 15 kg *per capita* per year by 2050, while China is projected to supply six times more. China's meat demand is projected to be 83 kg *per capita* per year by 2050.[57] The demand for meat is tied to economic growth, and global gross domestic product (GDP) is now in its fifth successive year of expansion at a rate of 4% or more (Figure 1.3).

Higher incomes in India and China have made hundreds of millions of people rich enough to afford meat and other foods. In 1985 the average Chinese consumer ate 20 kg (44 lb) of meat a year; now the figure is more than 50 kg.[58] Research conducted in China shows that in the last decade for each extra dollar of income, additional high-fat foods are purchased.[59] China's appetite for meat may be nearing satiation, but other countries are following behind: in developing countries as a whole, consumption of cereals has been flat since 1980, but demand for meat has doubled.

Most countries in Asia, Latin America, Northern Africa, the Middle East and the urban areas of sub-Saharan Africa are now experiencing an inexorable shift to the higher-fat Western diet, reflected in a large proportion of the population consuming over 30% of their energy from fat.[60] Economists speak of this effect as

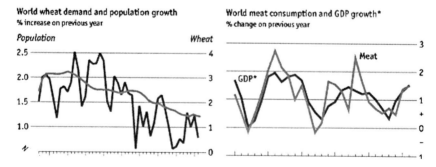

Figure 1.3   More people, more grain: more money, more meat. Source: Goldman Sachs, 2008.

one that shows how the decision-making demand pattern for food has changed, so that for the same income level the patterns of demand have changed significantly from earlier periods. What is evident is that humanity is increasingly consuming more and struggling to meet its food needs.

1.3.2 Misconsumption

In 2001, a glossary of terms issued by the World Trade Organization[61] defined food security, somewhat prejudicially, as: '... a concept which discourages opening the domestic market to foreign agricultural products on the principle that a country must be as self-sufficient as possible for its basic dietary needs.' Food security is not only a matter of food production or food supply; it is a function of availability, accessibility, stability of supply, affordability and the quality and safety of food. These factors include a broad spectrum of socio-economic issues with great influence on farmers and on the impoverished in particular. For some, food security is a term primarily associated with developing countries and for others it is synonymous with self-sufficiency.

The problem is not just the number of mouths to feed; it is the quantity of food that each mouth consumes when there are no natural constraints. As the world becomes richer, people eat too much and, in particular, too much of the wrong things—above all, meat.

The United Nations World Food Programme offers another persepctive: it affirms that the total surplus of the US alone could satisfy 'every empty stomach' in Africa, and France's leftovers could feed the Democratic Republic of Congo just as Italy's could feed Ethiopia's undernourished.[62] Food insecurity could therefore be seen as a consequence of unequal distribution and unbiased trade regulations.

Michael Pollan highlights that the food business once lamented what it called the problem of the 'fixed stomach'.[63] It appeared that demand for food, unlike other products, was inelastic, *i.e.* the amount fixed by the dimensions of the stomach itself and the variety constrained by tradition and habit. In the past few decades, however, American and European stomachs have become as elastic as balloons, and, with the newly prosperous Chinese and Indians switching to Western diets, much of the rest of the world is following suit. 'Today, Mexicans drink more Coca-Cola than milk,' Raj Patel reports in one of his latest books.[64] Paul Roberts highlights that in India '... obesity is now growing faster than either the government or traditional culture can respond',[65] and the demand for gastric bypass surgery is soaring.

Recently a research project entitled *UK Food Supply in the 21st Century: The New Dynamic*[66] has been carried out in the UK by a research team drawn from centres of excellence around the country. The research has ensured the gathering of opinion from around the supply networks concerned and from all over the UK. The results of the research outline the fact that Western societies, in particular, tend to take their food supply for granted. The thought of food shortages or supply chain disruptions hardly enters into people's minds. This is mainly because food has been cheaper in real terms, and more readily available, during the last few decades than probably at any time in history. As a consequence, food is given for granted, misused and often abused.

The number of overweight and obese people is definitely an increasing problem both in developed and in developing countries. Globally, there are roughly 50% more people who are overweight and obese (1.2 billion) than who are malnourished (860 million).[67] The reasons for overweight and obesity are complex. A high intake of energy-dense foods is just one of the many factors. While livestock products and fish are important in a nutritious diet, in many countries the consumption of livestock products, sugar and oil is significantly higher than what is required for human health.

Consumers' tastes changing towards more nutritious and more diversified diets must be kept in mind. Generally the poor eat vegetables, while the rich eat food that eats vegetables. Americans consume 120 kg of meat each per year; in the developing world people eat 28 kg of meat each per year.[68] As the trend continues, China and other countries undergoing the nutrition transition such as India, Thailand and Egypt will require ever greater amounts of animal feed, water and grazing land and are likely to have to import some of their feed and livestock products to meet future demand. This will tend to strengthen the negative environmental impact associated with the livestock sector. Rising demand as a result of the nutrition transition may best be described as a long-term underlying shift. It will continue to cause upward pressure on world food and feed crop prices and is likely to be influenced only marginally by global economic conditions.

While the risk of undernourishment is reduced with increasing supply of food—provided that access is ensured—the risk for overeating and wastage is likely to increase when food becomes more abundant in society. Availability of food, low prices, increased variety of flavours and larger portion sizes,[69] are all factors that induce us to eat more and more refined foods.

The future will tell if emerging environmental concerns and health problems related to a rich diet combined with appropriate policies may promote counter-trends, particularly in high-income countries.

1.4 THE INTERCONNECTION BETWEEN FOOD PRODUCTION, CONSUMPTION, TRANSPORTATION AND THE ENVIRONMENT

(Over)consumption is not the only aspect of the food chain to have multilevel implications, and in particular a strong impact on the environment. Agricultural production, processing, storage, transport, preparation and waste also have impacts on the environment in different ways relative to the life cycle of the food itself.

The environmental impact of food production and consumption is measured by the **climate foodprint**.[70] The concept of the climate foodprint falls within the scope of both the carbon footprint—which measures all greenhouse gases (GHGs) we individually produce through burning fossil fuels for electricity, heating and transportation and has units of tonnes (or kg) of CO_2 equivalent—and in a broader sense, the ecological footprint that measures how much land and water area a human population requires to produce the resource it consumes and to absorb its wastes, using prevailing technology.

In fact, the production and consumption of food generate an environmental impact both in terms of CO_2 emissions (carbon footprint) and in terms of demands on the Earth's ecosystems (ecological footprint). The carbon footprint can be subdivided into the so-called **fossil carbon footprint**, relating to the emissions of CO_2 into the atmosphere, and the so-called **biocarbon footprint**, which refers to the fact that CO_2 is actually absorbed from the atmosphere during the production process of a specific food. In other words, we must take into consideration, for example, that if on the one hand a certain quantity of CO_2 emissions is associated with the final consumption of a fruit, on the other hand the plant which provided that fruit has also absorbed CO_2 from the atmosphere through the process of photosynthesis.

To summarize, an average person who eats in accordance with a North American diet leaves a 26.8 m^2 ecological footprint and releases approximately 5.4 kg of CO_2 into the atmosphere each day. An average person who eats in accordance with a Mediterranean diet leaves a 12.3 m^2 ecological footprint and releases approximately 2.2 kg of CO_2 into the atmosphere each day.[71] The differences in ecological footprint and carbon footprint between the North American diet and the Mediterranean diet can be traced back mainly to the following factors:

- The quantity of food eaten, which is higher in the North American diet.
- The type of food eaten, which is mainly meat and sweet foods in the North American diet as opposed to the carbohydrates, fruit and vegetables that characterize the Mediterranean diet.
- The composition of the food eaten, which on the whole has a far greater calorie content in the North American diet compared to the Mediterranean diet, in terms of equivalent food type.

Not only do food production and consumption have an impact on the environment, food transport also significantly affects the environment and the evidence shows that more transport means more GHGs.[72] Within the transportation system itself, other factors are likely to be at least as important as distance, such as the mode of transport and the efficiency of the vehicles and systems. Ships are less carbon-intense than rail, which is in turn preferable to road transport, and trucks are preferable to planes.

Once one looks at the whole lifecycle of a product, transport for many food items is a relatively small component of total GHG emissions. A number of counter-intuitive case studies are widely quoted, where products transported further—even by air—can be less carbon-intensive than their 'local' counterparts. For example, tomatoes trucked to the UK from Spain in the winter are less carbon-intensive than those grown in heated British greenhouses. Kenyan roses air-freighted to the UK are considerably less carbon-intensive than those raised in hothouses and then trucked from the Netherlands.

Food transport is responsible for 2.5–3.5% of UK GHG emissions,[73] which is a slightly higher proportion than food manufacturing (2.2%) or food-related energy use in the home, such as cooking and chilling (2.1%), but less than, for example, total emissions associated with meat and dairy, which are responsible for around 8%. Such figures tell us that we need to compare the contribution transport makes to overall emissions in proportion compared with other emissions from within the food sector and beyond it. Issues such as seasonality and methods of production must also be kept in mind.

If it is true that food production, consumption and transportation have an impact on the environment, it is also true that, conversely, they are deeply affected by environmental degradation. For instance, global water scarcity may reduce crop yields by up to 12% and climate change may accelerate invasive pests of insects, diseases and weeds, reducing yields by an additional 2–6% worldwide. Continuing land degradation, particularly in Africa, may reduce yields by another 1–8%. Croplands may be swallowed up by urban sprawl, biofuels, cotton production and land degradation by 8–20% by 2050, and yields may become depressed by 5–25% due to pests, water scarcity and land degradation.[74]

Increased use of artificial fertilizers and pesticides, increased water use and cutting down of forests will result in massive decline in biodiversity. Already, nearly 80% of all endangered species are threatened due to agricultural expansion, and Europe has lost over 50% of its farmland birds during the last 25 years of intensification of farming practice. As a consequence, while new ways to produce extra food are researched, the rapidly changing climate is going to make the Earth a less efficient piece of farmland.

The 2 °C increase in average temperatures that is accepted as the likely minimum this century will be enough to cause major shifts in the seasons and in what crops work where. According to a research by the Intergovernmental Panel on Climate Change (IPCC),[75] just half a degree of average temperature increase will reduce the yield of India's wheat crop by 20%—and India is the world's second largest producer of wheat. If crops fail in India, we will feel the effects very quickly. The enormous price rises in staple foods in Europe in early 2008 were born in the great cereal lands of Brazil and India and in the rice paddies of south and east Asia.

In eastern and southern Asia, climate change is expected to affect rains, increase the frequency of droughts and raise average temperatures, threatening the availability of fresh water for agricultural production. In sub-Saharan Africa, arid and semi-arid areas are projected to increase significantly. In southern Africa, yields from rain-fed agriculture are expected to fall by up to 50% as early as 2020.

The impact of climate change on agriculture is likely to lead to a loss of stability in productivity and an overall decline in food production. Unless urgent action is taken, climate change will undoubtedly worsen global food security and dramatically increase the number of people facing hunger and malnutrition. Current estimates indicate that climate change could put a million more people at risk of hunger by 2020.[76]

The future of global food security is highly dependent on two important and interrelated factors. The first is the degree to which developing countries will succeed in raising agricultural productivity through technological change and effective natural resource management. The second is the degree to which the world will succeed in limiting climate change, while helping developing countries adapt to climate change and mitigate its effects. There are currently 500 million smallholder farmers worldwide who support around 2 billion people, or one-third of the world's population. Increasing their productivity is essential not only to secure the food and nutrition needs of these farmers, but also of the millions of people who depend on them.

1.5   THE FUTURE FOOD CHALLENGES

In all probability, 20 years from now the way food is produced, sold and consumed will have drastically changed. Projections of food supply and demand in fact suggest substantial transformations in consumption patterns, in technologies, in policies, and in international trade.

Future food security depends on developments in both supply and demand, but projections for these variables are cursed with uncertainty. On the demand side, population and economic growth are particularly subject to a high degree of uncertainty. Key uncertainties for future supply have to go with agricultural productivity and energy markets. In addition, developments are contingent on new policies being put in place.

Reilly and Willenbockel[77] review a number of exercises that have sought to explore alternative futures. Most have adopted one of a variety of types of scenario analysis, combining different assumptions about exogenous driving factors such as population, global economic growth and climate change with an endogenous economic modelling engine. The economic models are typically partial equilibrium models, where the prices of different types of food in different countries connected by trade are determined by specifying such things as how demand is driven by personal income, supply driven by likely changes in climate, *etc.* Assumptions are also made about endogenous productivity growth. Looking ahead, we can identify challenges to the food system and factors that will increase the demand for food as well as trends in the future food supply. Factors such as a growing population, incomes, urbanization, agriculture, and trade liberalization needs to be analysed together with the emerging role of food industry stakeholders (transnational food corporations, food retailers and food marketing) and the changing attitude and consumption patterns of consumers.

1.5.1   Growing Populations

According to the World Bank, each day 200 000 more people are added to the world food demand. The world's human population has increased near fourfold in the past 100 years; it is projected to increase from 6.7 billion (2006) to 9.2 billion by 2050, as shown in Figure 1.4.[78] It took only 12 years for the last billion to be added, a net increase of nearly 230 000 new people each day, who all need housing, food and other natural resources. The largest population increase is projected to occur in Asia, particularly in China, India and southeast Asia, accounting for about 60% or more of the world's population by 2050. The rate of population growth, however, is still relatively high in Central America, and highest in Central Africa and parts of West Africa. In relative numbers, Africa will experience the most rapid growth, over 70% faster than in Asia (annual growth of 2.4% *versus* 1.4% in Asia, compared to the global average of 1.3% and only 0.3% in many industrialized countries). In sub-Saharan Africa, the population is projected to increase from about 770 million to nearly 1.7 billion by 2050 (Figure 1.5).

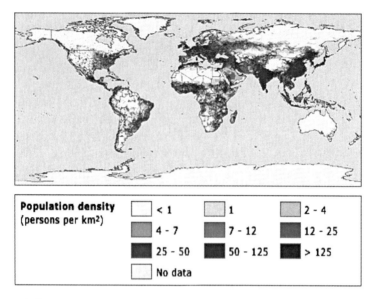

Figure 1.4 Population density. Source: http://www.worldometers.info/population/

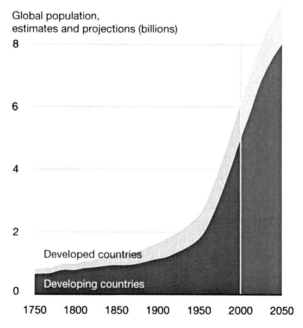

Figure 1.5 Human population growth in developed and developing countries (mid-range projection). Source: UN Population Division, 2007.

Estimates released by the World Bank in August 2008 show that in the developing world the number of people living in extreme poverty may be higher than previously thought. With a threshold of extreme poverty set at US$1.25 a day, there were 1.4 billion people living in extreme poverty in 2005. Each year, nearly 10 million people die of hunger and hunger-related diseases. While the proportion of underweight children below 5 years old decreased—from 33% in 1990 to 26% in 2006—the number of children in developing countries who were underweight still exceeded 140 million. Similarly, while the proportion of impoverished persons might have declined in many regions, their absolute number has not fallen in some regions as populations continue to rise.[79] There are huge regional differences in these trends. Globally, poverty rates have fallen from 52% in 1981 to 42% in 1990 and 26% in 2005.

Ageing populations and the rise in the number of single-parent families are likely to amplify the trend to smaller households, thereby affecting packaging requirements and the social circumstances of food consumption. In the EU, the share of single-person households is projected to climb from its current level of 30% to around 36% in 2015, and in Japan from 23% to 28% in 2010.

1.5.2 Income

Over the next three to four decades, global *per capita* income is projected to rise at a rate of over 2% per annum, with developing countries that are starting from a low base expected to rise at even higher rates.[80] Their economies are expected to expand at twice the rate of those in industrial countries. On a *per capita* basis, world real incomes may rise by 4.5 times by 2050.[81] Rising incomes mean higher-fat diets. In Mexico and Brazil, for example, where overweight used to be a sign of wealth, it now more often reflects poverty. Increased incomes or lower prices have led to the increased consumption of animal-based foods and processed foods. While those that are well educated can choose to adopt a healthy lifestyle, the poor have fewer food choices and more limited access to nutritional education.

Admitting that the pace and magnitude of economic growth cannot be predicted with a high degree of certainty, there is still a widespread view that the world economy, including most economies in Asia, Latin America and large parts of Africa, will continue to expand.[82]

Even if GDP projections are based on purchasing power parity calculations, the future effective demand for food and the mix of food items is extremely difficult to assess. It is, however, plausible that the economic factor is potentially a more forceful driver than population growth *per se.*

1.5.3 Urbanization

Essentially, almost all population growth over the coming decades will be urban. In 1900, just 10% of the world's population inhabited cities; today, that figure is over 50%. While urbanization will proceed very slowly in many industrial and transition countries, it will continue to grow unabated in those countries where the vast majority of the country is rural. This is already particularly evident in Sub-Saharan Africa (urbanization rate greater than 4%) and east Asia (urbanization rate greater than 3%). Urbanization has numerous consequences in that it leads to new and improved marketing (with greater access to modern mass media) and distribution infrastructure, attracts large supermarkets dominated by multinational corporations, and results in better transportation systems, thereby improving access to foreign suppliers and the importance of imports in the overall food supply.[83] This ultimately facilitates and results in the globalization of food consumption patterns.

Rapid urbanization has had, and will continue to have, a profound effect on food consumption patterns.[84] A higher calorific intake (cities offer a greater range of food choices), combined with a lower-energy expenditure in urban jobs (with a reduction of physical activity of the order of 10–15%) compared with rural work and more inactivity in leisure time, means that obesity and diabetes in developing countries are advancing more rapidly in cities than in rural areas.

Also, urbanization can affect food consumption by changes in dietary behaviour. This niche has been seized by the fast-food industry, by providing quick access to cheap takeaway meals. These meals also satisfy the consumers' demand for food high in salt, fat and sugar. Indeed, the most popular fast-food items, including hamburgers, pizzas and fried chicken, have 30% of their food energy as fat.[85] Thus, the major consequence of urbanization from a nutrition perspective is a profound shift towards higher food energy, more fats and oils and more animal protein from meat and dairy foods. This results in a diet that is lower in fibre, vitamins and minerals and higher in energy, total and saturated fat. Data from the China Health and Nutrition Survey, which found an increase in the consumption of animal products, observed that this increase was higher for urban residents compared with those living in the countryside.[86] Furthermore, intake of animal foods was greater for urban dwellers (178.2 g *per capita* per day) compared with rural dwellers (116.7 g *per capita* per day) in 1997.[86]

1.5.4   Agriculture

World agricultural production to 2015 is expected to grow at an average rate of around 1.8% a year; a slower pace than in preceding decades but fast enough to improve *per capita* food production as world population growth gradually loses momentum. The bulk of the expansion in production will be in developing countries, mainly the so-called BRIC countries—Brazil, Russia, India and China—largely due to the intensification of agriculture and a widespread use of agrochemical inputs.

Cereal demand projections are expected to be in the range of 2800–3200 Mt by 2050, an increase of 55–80% compared with today. Much of the future increase will be fed to animals to satisfy the demand for meat.[87] Today some 650 Mt of grain—nearly 40% of global production—is fed to livestock, and this may reach 1100 Mt by 2050.

Boosting world food production without gobbling up land and water requires technology to play a larger role in the next 40 years. Technology implies drip irrigation, no-till farming, more efficient ways to use fertilizers and kill pests, and also the enhancement of genetically modified (GM) crops that, for example, use less water.

The food price trauma of 2007–08 is persuading some countries to divert part of their wealth to subsidize food and support the development of new technologies so they can be self-sufficient and avoid future crises. But the demands of feeding 9 billion people in 2050 tell a different story: farming needs to be as efficient as possible. That requires markets and trade. Investing in agriculture is a boon; rejecting agricultural markets would be a disaster.

1.5.5 Trade Liberalization

Trade and investment in agro-food products are bound to play an essential role in feeding humanity in the coming decades. This will only come about, however, through further progress in reducing tariffs and export subsidies, and opening markets.

Trade liberalization is another important factor that has led to changes in food intake. Modifications in food supply have also radically altered the food environment and the choices that consumers may make. Reductions in the price of unhealthy foods, typically those that are calorie-rich, nutrient-poor and high in saturated fats and salt, compared with healthy foods, increased desirability and availability of unhealthy foods, worsening asymmetry between consumers and suppliers of foodstuffs, and growing urbanization and changes in lifestyle are all possible means by which trade liberalization can affect food consumption, especially among poor populations.[88]

Trade liberalization can affect the availability of certain foods by removal of barriers to foreign investment in food distribution. It can also enable foreign investment in other types of food retail; multinational fast-food outlets have made substantial investments in middle-income countries. The availability of processed food has risen in developing countries after foreign direct investment by multinational food companies.

Trade liberalization has enabled greater availability and affordability of highly processed, calorie-rich, nutrient-poor foods and animal products in developing countries, but more research is needed to better understand the relationship between trade policy and diets. Although data are available on the growth of processed food sales (29% annual growth in developing countries *versus* 7% in nations with high incomes[88]), evidence for consumption patterns of processed foods and their determinants in developing countries is still lacking.[89]

1.5.6 Factors Connected to Food Industry Stakeholders

Apart from the factors explained above, there are other aspects more deeply related to food industry stakeholders, *i.e.* the future role of transnational food corporations (franchises and manufacturers), retailers and food marketing, whose potential developments will deeply affect the future food scenario.

Transnational corporations (TNCs) are among the world's biggest economic institutions. A rough estimate suggests that the 300 largest TNCs own or control at least one-quarter of the entire world's productive assets, worth about US$5 trillion.[90] The total annual sales of TNCs are comparable to or greater than the yearly GDP of most countries.[91] In 1970, there were some 7000 parent TNCs, while today that number has jumped to 38 000, and 90% of them are based in the industrialized world.[92]

Table 1.4 presents the profiles of leading transnational food corporations. All these corporations have a strong global presence, with international sales close to or over 40% of their total sales for 2007. Nestlé's and McDonald's have had consistently high figures since 2000. The global share for McDonald's would be even higher if the total sales of international franchise units, and not just the corporate licensing revenues, could be included. For Kraft and PepsiCo as well, their total sales would be much higher than their net revenues.

Table 1.4 Profiles of leading transnational food corporations, 2007

Companies	Global Sales	Employees	Main Brands	# of Countries
Yum! (based on total company sales and franchise sales and employees)	$34 billion	1 million+	KFC, Taco Bell Pizza Hut, A&W	100+
McDonald's (based on total company sales and franchise sales and employees)	$47 billion	1.6 million	Has Owned: Chipotle Boston Market, Donatos Pizzeria	118
Kraft (based on net revenues)	$37 billion	104,000	Nabisco, Oscar Mayer, Post Cereal	155
Nestlé (based on total sales)	$90 billion	276,050	Nescafe, Hot Pockets, Crunch, Kit Kat	Almost world-wide
PepsiCo (based on total sales)	$93 billion	185,000	Pepsi, Frito-Lay, Gatorade, Tropicana	200

Source: Company annual reports and websites.

As countries undergo the process of development and globalization, they become exposed to Western products and brand diffusion, and the technologies that are needed to support these burgeoning forms of food consumption.

Retailers are the other big players in the food scenario. Their rapid growth was only possible because supermarkets expanded beyond their original markets, moving into small and poor countries, from urban to rural areas. In 2000, it was estimated that only five supermarket chains accounted for over 40% of food retail sales in the US and for over 60% in South America. This expansion of supermarkets now extends to east and southeast Asia as well as to eastern and central Europe.[93]

In those regions where supermarkets have made major inroads into the food retailing system, the entire food economy from farm to fork has been affected. Retailers impose their own rigorous uniformity and cosmetic standards on producers, driven by consumers' preferences for produce that appears uniform and aesthetically pleasing. Products that do not meet industry standards are sold at a loss or simply wasted. This induces many producers to use large, possibly excessive, amounts of pesticides, which results in pesticide residues making their way into purchasers.

If supermarkets bring the advantage of convenience, they also lead to an increased availability of cheaper, less healthy food, being large providers of processed, higher-fat, added-sugar and salt-laden foods, especially in developing countries, affecting the diet and the consumption patterns.

1.6   AN ALMOST NEGLECTED ISSUE: FOOD WASTE

Food chain sustainability, health concerns, intelligent consumer behaviour, food safety, ethical and environmental issues, climate change and catastrophic events, technological advances and proper food policies are set to become essential features of the future food scenario. The future impact of some of these factors is so unpredictable that it is difficult to see how they can realistically be incorporated into any quantitative models, other than by including some essentially arbitrary tolerance limits in calculating necessary food supplies.

While public interest in sustainability continues to rise and consumer attitudes are mainly positive, behavioural patterns are not always consistent with these attitudes. The selection of foods that are acceptable to an individual increasingly takes place in a context where availability is substantially influenced by the food industry and food retailers.

While a minimum food consumption level is needed for survival, the levels of consumption seen by modern Western culture vastly exceed those levels, compromising natural resources and at the same time creating huge amounts of waste. Juliet Schor in *The Overspent American*[94] suggests that: 'Consumer satisfaction depends less on what a person has in an absolute sense than on socially formed aspirations and expectations'. Similarly, the anthropologist Willett Kempton[95] affirms that from an environmental perspective, a problem with consumption to display social status is that status is always relative, generating an unending spiral of increasing consumption, display and re-comparison.

It is important to add the food waste issue to the list of factors. Surprisingly, most of the discussion on the food supply chain has so far ignored food waste. Despite a recent interest in the topic, food waste has largely not been taken into consideration. The interest is demonstrated by the flourishing of food waste studies (in 2010 the European DG Environment published the most comprehensive study on bio-waste[96]) and recent books and articles (the two most comprehensive books seem to be *Waste, Uncovering the Global Food Scandal*[97] by Tristram Stuart and *American Wasteland* by Jonathan Bloom[98]) and the recent State of the World 2011 Report launched by the Worldwatch Institute, which spotlights successful agricultural innovations and unearths major successes in preventing food waste.

Even international organizations such as FAO and the United Nations Environment Programme (UNEP) have dedicated little attention to food waste. Waste is a deeply inconvenient topic: it is a proof that the food system is uneven, that the mechanism behind the food supply chain aims at increasing waste (from surplus and overproduction to overconsumption) and that food waste goes unnoticed simply because we in the Western world are too used to having plenty of food at our disposal.[99]

Food waste originates along the whole food supply chain and far too many stakeholders are involved. Quantifying food waste at a national level is difficult because traditional methods rely on structured interviews, measurement of plate waste, direct examination of garbage and application of inferential methods using waste factors measured in sample populations and applied across the food system.[100] All of these methods are difficult to apply.

Wasting food is costly, damages the environment and, in a world where a billion people are hungry, is ethically and morally deeply unfair. In the next chapters we will not only illustrate how food waste originates and how all the stakeholders in the food chain contribute to it, but we will also present interesting data from all over the world explaining how the current polices on food waste are integrated into the broader food policies. We will also introduce and explain the functioning of some initiatives launched all over the world whose aim is to give food waste a second life and possibly transform it into a resource.

REFERENCES

1. P. Roberts, *The End of Food*, Bloomsbury, 2008, p. 4.
2. K. Kiple and K. C. Ornelas, *The Cambridge World History of Food*, Cambridge University Press, 2000, pp. 61–62.
3. R. Shatin, *The Transition from Food-gathering to Food-production in Evolution and Disease*, Vitalst Zivilisationskr, 1967, pp. 104–7.
4. M. N. Cohen, *Health and the Rise of Civilization*, London, Yale University Press, 1989.
5. C. Lévi-Strauss, *Le Totémisme aujourd'hui*, Puf, Paris, 1962.
6. S. B. Eaton, *Humans, Lipids and Evolution*, Lipids, 1992, p. 814.
7. http://history-world.org/agriculture.htm
8. http://www.enotes.com/peoples-chronology/year-prehistory
9. K. Kiple and K. Ornelas, *The Cambridge World History of Food*, Cambridge, UK, Cambridge University Press, 2000.
10. United Nations, *The World at Six Billion Report*, 1999 available at www.un.org/esa/population/publications/sixbillion/sixbilparti.pdf
11. R. Tannahill, *Food in History*, New York, Three Rivers Press, 1995, p. 72.
12. http://www.historylink101.com/lessons/farm-city/china1.htm
13. J. Flandrin and M. Montanari, *Food: A Culinary History*, New York: Penguin, 2000.

14. M. L. King, *Western Civilization: A Social and Cultural History*, v. 1 (Documents Set, eds. Arlene M. W. Sindelar and Mary E. Chalmers), New Jersey: Prentice Hall, 2000, pp. 147–48.
15. T. Hugh, *World History: The Story of Mankind from Prehistory to the Present*, New York, Harper Collins, 1996.
16. J. M. Roberts, *A History of Europe*, New York: Allen Lane/Penguin, 1997.
17. F. Braudel, *The Wheels of Commerce. Civilization and Capitalism: 15th-18th Century*, Volume 2, 1967, p. 61.
18. T. Malthus, *An Essay on the Principle of Population*, London, printed for J. Johnson, in St. Paul's Church-Yard 1798 available at http://www.esp.org/books/malthus/population/malthus.pdf
19. O. T. Solbrig and D. J. Solbrig, *So Shall You Reap: Farming and Crops in Human Affairs*, Washington, D.C., Island Press, 1994.
20. C. Ponting, *A Green History of the World—The Environment and the Collapse of Great Civilizations*, Penguin, 1993.
21. J. R. Kloppenburg, D. L. Kleinman and J. Handelsman, *First the Seed*, University of Wisconsin Press, 2004, p. 73.
22. http://www.localhistories.org/20thcent.html
23. T. Hugh, *World History: The Story of Mankind from Prehistory to the Present*, New York: HarperCollins, 1996.
24. IFPRI, *Green Revolution: A Curse or Blessing?*, Washington D.C., 2002, available at http://www.ifpri.org/sites/default/files/pubs/pubs/ib/ib11.pdf
25. L. K. Dyson, *American Cuisine in the 20th Century-A Century of Change in America's Eating Patterns*, Food Review, Volume 23, Issue 1, 2000. Available at http://www.ers.usda.gov/publications/foodreview/jan2000/frjan2000a.pdf
26. D. E. Bowers, Cooking Trends Echo Changing Roles of Women, Food Review, Volume 23, Issue 1, Jan 2000 available at http://findarticles.com/p/articles/mi_m3765/is_1_23/ai_63543001/
27. M. Pollan, *The Omnivore's Dilemma*, Penguin, 2006.
28. H. Friedmann, *The International Political Economy of Food: A Global Crisis* in International Journal of Health Services, Vol. 25(3), 1995, p. 511–538.
29. T. Lang and M. Heasman, *Food Wars. The Global Battle for Mouths, Mind and Markets*, Earthscan, London, 2004, p. 364.

30. P. J. Ericksen, *Conceptualizing Food Systems for Global Environmental Change Research*, Global Environmental Change, 2007.
31. J. Pretty J and H. Ward, *Social Capital and the Environment*, World Development 29 (2), 2001, p. 209–227.
32. P. A. Matson, W. J. Parton, A. G. Power and M. Swift, *Agricultural Intensification and Ecosystem Properties*, Science 1997, p. 504–509.
33. T. Reardon and J. A. Berdegué, *The Rapid Rise of Supermarkets in Latin America: Challenges and Opportunities for Development*, in Development Policy Review 20(4), 2002, p. 371–88.
34. M. A. Mendez, C. A. Montero and B. M. Popkin, *Is Obesity fuelling Inequities in Health in the Developing World?*, 2004, University of California Manuscript, Chapel Hill, NC.
35. J. Kennedy, V. Jackson, I. S. Blair, D. A. McDowell, C. Cowan and D. J. Bolton, *Food Safety Knowledge of Consumers and the Microbiological and Temperature Status of their Refrigerators* in Journal of Food Protection, 68(7), 2005, p. 1421–1430.
36. J. Lundqvist, C. de Fraiture and D. Molden, *Saving Water: From Field to Fork—Curbing Losses and Wastage in the Food Chain*, SIWI Policy Brief, SIWI, 2008.
37. http://www.earthtimes.org/articles/news/281189,world-population-will-excede-9-billion-by-2050.html
38. L. Brown, *Could Food Shortages Bring Down Civilization?*, Scientific American, May 2009.
39. FAO, *How to Feed the World in 2050*, Report 2009 available at www.fao.org/fileadmin/.../How_to_Feed_the_World_in_2050.pdf
40. FAO, *The State of Food Insecurity in the World 2009. Economic crises—impacts and lessons learned*, 2009, available at http://www.fao.org/docrep/012/i0876e/i0876e00.htm
41. J. von Braun, *The World Food Situation—New Driving Forces and Required Actions*, IFPRI Publication, Washington D.C., December 2007.
42. J. Lundqvist, C. de Fraiture and D. Molden, *Saving Water: From Field to Fork—Curbing Losses and Wastage in the Food Chain*, SIWI Policy Brief, SIWI, 2008.
43. www.ft.com/worldfoodoct2009

44. FAO, *The State of Food and Agriculture*, Rome, 2008.
45. UNDP, *Human Development Report*, Washington D.C., 1998.
46. http://www.worldrevolution.org/projects/globalissuesoverview/overview2/BriefPeace.htm
47. World Development Bank Indicators, *Poverty data*, a supplement to World Development Indicators, Washington D.C., 2008.
48. United Nations Development Programme (UNDP), *Human Development Report 1998*, Overview.
49. FAOSTAT, *Statistical database 2007*, available at http://faostat.fao.org
50. L. Serra-Majem, D. Maclean, L. Ribas, D. Brulé, W. Sekula, Garcia-Closas, A. Yngve, M. Lalonde and A. Petrasovits, *Comparative Analysis of Nutrition Data from National, Household, and Individual Levels: Results from a WHO-CNIDI Collaborative Project in Canada, Finland, Poland, and Spain*, in Journal of Epidemiology and Community Health, 2003, p. 74–80.
51. FAOSTAT, *Updating the Minimum Dietary Energy Requirements*, Rome, October 2008 available at http://www.fao.org/fileadmin/templates/ess/documents/food_security_statistics/metadata/undernourishment_methodology.pdf
52. Swaminathan Research Foundation, *Atlas of the Sustainability of Food Security in India*, MSSF, 2002.
53. V. Smil, *Feeding the world. A challenge for the Twenty-first Century*, Cambridge, MA: MIT Press, 2000.
54. FAOSTAT, *Statistical Database 2007*, available at http://faostat.fao.org
55. N. Alexandratos, *World Agriculture: Towards 2030/50*, Interim Report. A FAO perspective. London, UK: Earthscan; Rome, Italy: FAO, 2006.
56. V. Smil, *Feeding the World. A Challenge for the Twenty-first Century*, Cambridge, MA: MIT Press, 2000.
57. C. de Fraiture, L. Karlber, J. Rockstrom, *Can Rainfed Agriculture Feed the World? An Assessment of Potentials and Risk*, In Wani *et al.*, eds. *Rainfed Agriculture: Unlocking the Potential*. CAB International, 2008, pp. 124–132.
58. www.economist.com/node/10250420

59. X. Guo, T. Mroz, B. M. Popkin and F. Zhai, *Structural Changes in the Impact of Income on Food Consumption in China, 1989–93*, Econ. Dev. Cult. Change, 2000, p. 740.

60. B. M. Popkin, *Global Nutrition Dynamics: the World is shifting rapidly Toward a Diet linked with Non-communicable Diseases*, Am. J. Clin. Nutr. 84, 2009, p. 298.

61. http://www.wto.org/english/thewto_e/whatis_e/tif_e/understanding_text_e.pdf

62. World Food Programme, *Comprehensive Food Security and Vulnerability Analysis (CFSVA): An External Review of WFP Guidance and Practice*, May 2006.

63. M. Pollan, *In Defense of Food: an Eater's Manifesto*, Penguin 2009.

64. R. Patel, *Stuffed and Starved. The Hidden Battle for the World Food System,* Portobello Books Ltd, 2007.

65. P. Roberts, *The End of Food*, New York: Houghton Mifflin, 2008.

66. Chatham House, *Food Supply in the 21st Century: rethinking UK Strategy*, Report, 9th February 2009.

67. J. Lundqvist, C. de Fraiture and D. Molden, *Saving water: From Field to Fork-Curbing losses and wastage in the food chain*, SIWI Policy Brief, Swedish International Water Institute: Stockholm, 2008.

68. A. Bowman, K. Mueller and M. Smith, *Increased Animal Waste Production from CAFOs: Potential Implications for Public and Environmental Health*, Omaha, Neb.: Nebraska Center for Rural Health Research, Occasional Paper #2, 2000.

69. H. A. Raynor and L. H. Epstein, *Dietary Variety, Energy Regulation and Obesity*, in Psychological Bulletin, 127, 2001, pp. 325–241.

70. J. Edwards, J. Kleinschmit and H. Schoonover, *Identifying our Climate 'Foodprint' Assessing and Reducing the Global Warming Impacts of Food and Agriculture in the U.S.*, IATP Publication, 2009.

71. A. Kouris-Blazos, C. Gnardellis, M. L. Wahlqvist, D. Trichopoulos, W. Lukito and A. Trichopoulou, *Are the Advantages of the Mediterranean Diet transferable to other*

Populations? A cohort Study in Melbourne, Australia, Br J Nutr, 1999.

72. T. Lang, *Local/Global (Food Miles)*, Lecture, Slow Food, Bra, Italy, 19 May 2006.
73. Food Ethics Council, *Food Distribution: an Ethical Agenda*, 2008 available at http://www.ifr.ac.uk/waste/Reports/Food-Ethics-Countil-fooddistribution.pdf
74. C. Nellemann, M. MacDevette, T. Manders, B. Eickhout, B. Svihus, A. G. Prins and B. P. Kaltenborn, *The Environmental Food Crisis—The Environment's Role in averting Future Food Crises*, A UNEP Rapid Response Assessment, 2009.
75. R. T. Watson, M. C. Zinyowera and R. H. Moss, *The Regional Impacts of Climate Change*, IPCC, Cambridge University Press, UK, 1997, p. 517.
76. http://www.countercurrents.org/nwanze111109.htm
77. M. Reilly, D. Willenbockel, *Managing Uncertainty: a Review of Food System Scenario Analysis and Modeling*, Philos Trans R Soc Lond B Biol Sci, 2010, Sep 27, pp. 349–63.
78. *UN Population Division Policy Brief* No. 2007/1, March 2007.
79. http://www.undp.org/publications/annualreport2008/
80. S. Du, T. A. Mroz, F. Zhai and B. M. Popkin, *Rapid Income Growth adversely affects Diet Quality in China*, Soc Sci Med 59, 2004, pp. 1505–1515.
81. J. Sachs, *Common Wealth: Economics for a Crowded Planet*, The Penguin Press, New York, 2008.
82. T. Lind and B. Malmberg, *Demographically based Global Income Forecasts up to year 2050*, International Journal of Forecasting, Vol. 23, 2007, pp. 553–567.
83. C. Hawkes, *Uneven dietary Development: linking the Policies and Processes of Globalization with the Nutrition Transition, Obesity and Diet-Related Chronic Diseases*, in Global Health 2, 2006.
84. B. M. Popkin, *Urbanization, Lifestyle Changes and the Nutrition Transition*, World Dev. 27, 1905–1916, 1999.
85. V. Smil, *Feeding the World. A Challenge for the Twenty-first Century*, Cambridge, MA: MIT Press, 2000.

86. F. Zhai, H. Wang, S. Du, Y. He, Z. Wang, K. Ge and B. M. Popkin, *Prospective Study on Nutrition Transition in China*, Nutr. Rev. 67 (Suppl. 1), S56–S61, 2009.

87. C. de Fraiture, D. Wichelns, E. Kemp Benedict and J. Rockstrom, *Scenarios on Water for Food and Environment*, in Water for Food, Water for Life: A Comprehensive Assessment of Water Management in Agriculture', Chapter 3, Earthscan, London and International Water Management Institute, Colombo, 2007.

88. A. M. Thow, *Trade Liberalisation and the Nutrition Transition: Mapping the Pathways for Public Health Nutritionists*, Research Paper, Cambridge Journals Online, volume 12, issue 11, 2009, available at http://journals.cambridge.org/action/displayAbstract?fromPage=online&aid=6324508

89. C. Hawkes, IFPRI, *The Role of Foreign Direct Investment in the Nutrition Transition*, Cambridge Journals Online, Volume 8, Issue 04, 2007.

90. www.unctad.org/en/docs/iteiitv2n2a6_en.pdf

91. *Everybody's Favourite Monsters*, The Economist, Survey of Multinationals, 27 March 1993.

92. F. Clairmonte and J. Cavanagh, *The World in Their Web— The Dynamics of Multinationals*, Zed Press, London, 1981, pp. 5–6.

93. L. Dries, T. Reardon and J. F. M. Swinnen, *The Rapid Rise of Supermarkets in Central and Eastern Europe: Implications for the Agrifood Sector and Rural Development*, Development Policy Review, Vol. 22, No. 5, pp. 525–556, September 2004. Available at SSRN: http://ssrn.com/abstract=584046

94. J. Schor, *The Overspent American: Why We Want What We Don't Need*, HarperCollins Publishers, 1999.

95. W. Kempton, *Cognitive Anthropology and the Environment*, in New Directions in Anthropology & Environment, Walnut Creek: AltaMira Press, 2001, pp. 49–51.

96. http://ec.europa.eu/environment/waste/compost/developments.htm

97. T. Stuart, *Uncovering the Global Food Scandal*, Penguin Press, 2009.

98. J. Bloom, *American Wasteland*, Da Capo Lifelong Books, 2010.

99. A. Segrè, *Dalla Fame alla Sazietà, dalle Eccedenze allo Spreco Inutile*, Georgofili—Atti della Accademia dei Georgofili. Fame e spreco alimentare. Trasformare le eccedenze in risorse a fini solidali. Il caso Last Minute Market. Firenze, Accademia dei Georgofili. 16 ottobre 2006. (vol. 3, p. 45) Firenze, Italy, Accademia dei Georgofili, 2004.

100. M. Griffin, M. J. Sobal and T. A. Lyson, *An Analysis of a Community Food Waste Stream*, Agric Hum Values, 2009, p. 67.

CHAPTER 2

Why Food Waste is Everywhere

The waste of plenty is the resource of scarcity.

Thomas Love Peacock (1785–1866)

2.1 INTRODUCTION

Food waste and its consequences go largely unnoticed. Food waste is common and unseen. Abandoned harvests, supermarkets and restaurants where abundance always reigns are just a symbol of the unseen food waste. Whether it comes from individual choice, a business mistake or government policy, food waste is generated from decisions made somewhere in the long food supply chain.

Although things are slowly changing—more scientific research is being conducted on this topic, print and television journalists are writing pieces on how to reduce food waste, some restaurants are reducing portion sizes, new 'frugal' blogs that dot the web are growing in number like 'premeditated leftovers', 'stuff on rice', 'not made of money', 'myzerowaste'—there is still a long way to go and we still live in a waste culture which developed in a historically short period of time.

Transforming Food Waste into a Resource
By Andrea Segrè and Silvia Gaiani
© Andrea Segrè and Silvia Gaiani, 2012
Published by the Royal Society of Chemistry, www.rsc.org

2.2 HOW TO DEFINE FOOD WASTE

Defining food waste is not easy. Although waste is formally defined in different legal jurisdictions, definitions relate to particular points of origin and are often framed in relation to specific environmental controls. Since there are several definitions of waste, equally many definitions of food waste exist. As a consequence, professional bodies, including international organizations, state governments and secretariats, may formally have their own definitions.

Although food waste occurs at different points in the food supply chain, it is most readily defined at the retail and consumer stages, where outputs of the agricultural system are self-evidently 'food' for human consumption.

Unlike most other commodity flows, food is biological material subject to degradation, and different foodstuffs have different nutritional values. There are also moral and economic dimensions: the extent to which available food crops are used to meet global human needs directly, or diverted into feeding livestock, other by-products or the production of biofuels or biomaterials.

Below are three definitions referred to food waste:

1. Wholesome edible material intended for human consumption, arising at any point in the food supply chain that is instead discarded, lost, degraded or consumed by pests.[1]
2. As (1), but including edible material that is intentionally fed to animals or is a by-product of food processing diverted away from human food.[2]
3. As definitions (1) and (2) but including overnutrition—the gap between the energy value of consumed food *per capita* and the energy value of food needed *per capita*.[3]

The first two definitions are considered to be most relevant.

Broadly speaking, food waste is composed of raw or cooked food materials and includes food loss before, during or after meal preparation in the household, as well as food discarded in the process of manufacturing, distribution, retail and food service activities. It includes materials such as vegetable peelings, meat trimmings, and spoiled or excess ingredients or prepared food as well as bones, carcasses and organs.

The Organisation for Economic Co-operation and Development (OECD) defines waste as:

> ... materials that are not prime products (*i.e.* products produced for the market) for which the generator has no further use for his own purpose of production, transformation or consumption, and which he discards, or intends or is required to discard. Waste may be generated during the extraction of raw materials, during the processing of raw materials to intermediate and final products, during the consumption of final products, and during any other human activity.[4]

In 1975, the European Commission, the executive branch of the European Union (EU), legally defined waste for countries in the Union in the EU Council Directive Waste 75/442/EEC as: 'any substance or object which the holder disposes of or is required to dispose of pursuant to the provisions of national law in force'.[5] This directive was amended in 1991[6] (by 91/156), with the addition of 'categories of waste' and the omission of any reference to national law. Types of waste are categorized by how they occur, and some categories appear specific to certain waste types; 'products whose date for appropriate use has expired' targets food waste, with 'date' referring to the expiry date of a food.

The United States Environmental Protection Agency (US EPA) defines food waste for the US as being:

> ... Uneaten food and food preparation wastes from residences and commercial establishments such as grocery stores, restaurants, and produce stands, institutional cafeterias and kitchens, and industrial sources like employee lunchrooms.

Although the EPA is a nationwide agency, states are free to define food waste individually, according to policies, preference and other definitions, though many choose not to.[7]

Overall, the definition of food waste can vary in many ways, including, but not limited to: what food waste consists of, how food waste is produced, and where/what it is discarded from/generated by. The definition can be varied and complicated by other issues. Certain groups do not consider (or have traditionally not considered) food waste to be a waste material, because of its applications; some definitions of what food waste consists of are based on other waste definitions (*e.g.* agricultural waste), and which materials do not meet their definitions.

The Waste & Resources Action Programme (WRAP), which works to help businesses and individuals in the UK reap the benefits of reducing waste, makes a further distinction in its report *Household Food and Drink Waste in the UK*,[8] between edible and inedible food waste (Table 2.1).

Some scientists make a further distinction between 'absolute waste', *i.e.* obvious waste which involves the actual destruction or deterioration of food, and 'waste in the relative sense', involving the use of food in ways which do not return a maximum quantity of balanced nutrients for human consumption (Table 2.2).[9]

Food deterioration occurs from microbial contamination or from rotting as a consequence of overproduction, storage problems or improper preparation. Food waste also occurs through food use that returns little nutritional value, like overprocessing and overconsumption.

In general, food waste may thus be defined as:

> ... a less than maximum use of nutrients for human consumption...food waste is the destruction or deterioration of food or the use of crops, livestock and livestock products in ways which return relatively little human food value.[10]

It is therefore clear that there is not a single and commonly agreed definition of food waste.

Table 2.1 Edible and inedible food waste

Edible food waste	
Avoidable food waste	Food that is thrown away that was, at some point prior to disposal, edible (*e.g.* slices of bread, apples, meat)
Possibly avoidable food waste	Food that some people eat and others do not (*e.g.* bread crusts, potato skins)
Inedible food waste	
Unavoidable food waste	Waste arising from food preparation that is not, and has not, been edible under normal circumstances (*e.g.* bones, egg shells, pineapple skins)

Source: Based on *Household Food and Drink Waste in the UK*, WRAP, 2009.

Table 2.2 Food recoverable/not recoverable for human consumption[a]

Recoverable for human consumption	Not recoverable for human consumption
Edible crops remaining in farmers' fields after harvest	Livestock condemned at slaughter because of disease
Produce rejected because of market 'cosmetics' (blemishes, misshapen, *etc.*)	Diseased or otherwise unsafe produce
Unsold fresh produce from wholesalers and farmers' markets	Spoiled perishable food, including meat, dairy, and prepared items
Surplus perishable food from restaurants, cafeterias, caterers, grocery stores, and other food service establishments	Plate waste from food service establishments
Packaged foods from grocery stores, including overstocked items, dented cans, and seasonal items	Losses of edible portions associated with processing, such as skin and fat from meat and poultry, and peels from produce

Source: http://www.ers.usda.gov/Publications/FoodReview/Jan1997/Jan97a.pdf
[a]Foods are considered inedible when their quality deteriorates until they become unhealthy or noxious.

2.3 HISTORICAL AND CULTURAL CHANGES IN FOOD WASTE

Historical transformations have changed the type and amount of food waste generated over the years. Hunter-gatherer cultures discarded bones as their primary food waste. The development of agriculture added more plant materials to the food waste stream. Industrialized agriculture increased organic waste by-products from large-scale food processing. Increased population growth and urbanization multiplied and concentrated the amount of food waste, which was increasingly dumped as the cities that generated waste became located farther from agricultural areas.

Historical shifts have also occurred in the concept of food waste. Material prosperity has reduced the economic necessity for food conservation and re-use, and conspicuous consumption and disposal are demonstrations of social status.

When the Food and Agriculture Organization of the United Nations (FAO) was established in 1945, it had reduction of food losses within its mandate. By 1974, the first World Food Conference identified reduction of post-harvest losses as part of the solution in addressing world hunger.[11] At this time, an overall estimate for post-harvest losses of 15% had been suggested, and it was resolved to bring about a 50% reduction by 1985. Consequently, the FAO established the Special Action Programme for the Prevention of Food Losses.[12] The main focus was initially on reducing losses of durable grain, but it was later broadened to cover roots and tubers, fresh fruits and vegetables. Poor adoption rates for interventions led to the recognition that a purely technical focus was inadequate for solving problems within the sector and a more holistic approach was developed.[13]

There is no account of progress towards the 1985 post-harvest loss reduction target, and only recently Prof. Jan Lundqvist and the Global Action Against Food Waste[14] called for action to reduce food waste, advocating a 50% reduction in post-harvest losses to be achieved by 2025.

Food in post-industrial societies is inexpensive relative to total income, and wasting food is increasingly accepted. Societies with abundant food supplies often consider reusing leftover foods as inconvenient, while less food-rich societies regard food re-use as imperative. Specific parts of animals and plants considered edible in some cultures are considered inedible in others. Animal parts viewed as waste may include bones or shells, skins or scales, fat, blood, intestines, brains, eyes, and stomachs. Plant parts viewed as waste may include cores, seeds, stems, outer leaves, shells, rinds, husks or peels.

Cultural variations exist in what is considered garbage, and understanding cultural food rules is crucial in examining food waste.[15] Cultural differences in beliefs about what is edible versus inedible exist more often for animal foods than for plant foods. For example, intestines and other internal organs are considered delicacies in China but are discarded as offal in many Western countries. Animal fats are consumed or used as fuel in societies like the Inuit, but in post-industrial nations fat is often trimmed and discarded to reduce calorie intake. Blood is an ingredient in dishes like black pudding in Britain but is discarded in many other societies.

Some populations, like the Cajuns, an ethnic group mainly living in Louisiana, are proud of their efficient use of all parts of a slaughtered animal, and they claim to use 'everything except the squeal' of hogs.

Moral values in most cultures admonish food waste. In Japanese the term 'mottainai'[16] refers to 'a sense of regret concerning waste when the intrinsic value of an object or resource is not properly utilized'. It is a compound word: 'mottai' refers to the intrinsic dignity or sacredness of a material entity, while 'nai' indicates an absence or lack. The expression 'mottainai' can be uttered alone as an exclamation when something useful, such as food or time, is wasted. In addition to its primary sense of 'wasteful', the word is also used to mean 'impious; irreverent' or 'more than one deserves'.[17]

Agricultural societies often feed plant food wastes to animals, and many industrial societies process by-products of animal slaughter into livestock feed. Such practices recycle undesired by-products into edible foods and minimize actual food waste. Some societies accept the waste of less-desirable portions of animals and plants as a sign that they have attained a state of affluence and can afford to consume only high-quality items.

It is evident that there is a constantly changing relationship between waste and value. Food waste keeps tracking up and down between these two extremes. Its status is not stable, but flexible: some communities may discard orange peel, while for others it might be part of their breakfast. In brief, food waste is the generated outcome of specific value consumption. What is common to all cultures is that food waste is generally considered unclean, and that some food waste is generated for reasons of hygiene.

Despite its nutritional role in sustaining health and wellbeing, food waste is also a means of expressing one's social identity. John Scanlan[18] notes that producing waste is a core feature of modern life. Others have noted the ambivalent relationship we have with the waste we produce. C. T. Anderson[19] notes that we are both creator and agent of its disposal. It is our ambivalence towards waste, coupled with its ubiquity, that allows waste materials to be described so variously: negatively as garbage, trash and rubbish, or more positively as by-products, leftovers, offcuts, trimmings, and recycled material.[20]

Regardless of consumption and disposal practices, what is certain is that the growing world population, urbanization, the increased demand for more processed foods, a decline in the agricultural sector and many other factors have increased food waste (Figure 2.1).

A sort of evolution and story of food waste is outlined and presented in Box 2.1.

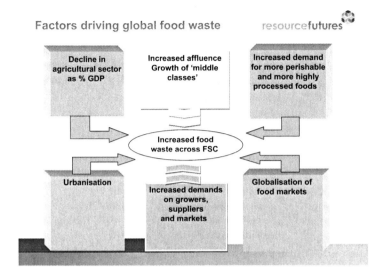

Figure 2.1 Factors driving Global Food Waste. Source: J. Parfitt, Food Supply, Consumption & Waste, Powerpoint Presentation, 7 September 2010, Resource Futures.

BOX 2.1 Food Waste Timeline

• **400 BC** The first municipal dump is established in ancient Athens.

• **AD 200** The first sanitation force is created by the Romans; teams of two men walk along the streets, pick up garbage and throw it in a wagon.

• **1388** The English Parliament bans dumping of waste and food waste in ditches and public waterways.

• **1551** The first recorded use of packaging: German paper-maker Andreas Bernhart begins placing his paper in wrappers labelled with his name and address.

• **1657** New Amsterdam (now Manhattan) passes a law against casting food waste in the streets.

• **1710** Colonists in Virginia commonly bury their trash. Holes are filled with building debris, broken glass or ceramic objects, oyster shells and animal bones. They also throw away hundreds of suits of armour that were sent to protect colonists from arrows of native inhabitants.

- **1792** Benjamin Franklin uses slaves to carry Philadelphia's waste and food waste downstream.
- **1800** Visitors describe New York City as a 'nasal disaster, where some streets smell like bad eggs dissolved in ammonia'.
- **1810** The tin can is patented in London by Peter Durand.
- **1834** Charleston, West Virginia, enacts a law protecting vultures from hunters. The birds help eat the city's garbage.
- **1866** New York City's Metropolitan Board of Health declares war on garbage, forbidding the 'throwing of dead animals, garbage or ashes into the streets'.
- **1874** The organized incineration of collected trash begins in Nottingham, England.
- **1879** Frank Woolworth opens the first five and dime store in Utica, New York. He pioneers the idea of displaying goods on open counters so customers can see and feel merchandise (a practice that later makes larger, theft-proof packaging necessary).
- **1879** 'Thither were brought the dead dogs and cats, the kitchen garbage and the like, and duly dumped. This festering, rotten mess were picked over by rag pickers and wallowed over by pigs, pigs and humans contesting for a living from it, and as the heaps increased, the odours increased also, and the mass lay corrupting under a tropical sun, dispersing the pestilential fumes where the winds carried them.'—Minister describing the New Orleans dump to the American Public Health Association.
- **1885** The first garbage incinerator in the US is built on Governors Island in New York Harbor.
- **1885–1908** 180 garbage incinerators are built in the US.
- **1889** 'Appropriate places for (refuse) are becoming scarcer year by year, and the question as to some other method of disposal ... must soon confront us. Already the inhabitants in proximity to the public dumps are beginning to complain.'—Health Officer's report, Washington, DC.
- **1892** Beer bottles now sport a metal cap to prevent spoilage.
- **1893** 'The means resorted to by a large number of citizens to get rid of their garbage and avoid paying for its collection would be very amusing were it not such a menace to public health. Some burn it, while others wrap it up in paper and carry it on their way to work and drop it when unobserved, or throw it

into vacant lots or into the river.'—Boston Sanitary Committee.

• **1894** The citizens of Alexandria, Virginia are disgusted by the sight of barge loads of garbage floating down the Potomac river from Washington, DC. They take to sinking the barges upriver from their community.

• **1897** The first recycling centre is established in New York City.

• **1900** American cities begin to estimate and record collected wastes. According to one estimate, each American produces annually 80–100 lb of food waste, 50–100 lb of rubbish and 300–1200 lb of wood or coal ash—up to 1400 lb per person altogether.

• **1900** Small and medium-sized towns build piggeries, where swine are fed fresh or cooked garbage. One expert estimates that 75 pigs can eat 1 t of refuse per day.

• **1902** A survey of 161 cities by the Massachusetts Institute of Technology finds that 79% of them provide regular collection of refuse.

• **1905** New York City begins using a garbage incinerator to generate electricity to light the Williamsburg Bridge.

• **1909** 102 of 180 incinerators built since 1885 are abandoned or dismantled. Many had been inadequately built or run. Also, America's abundant land and widely spaced population made dumping garbage cheaper and more practical.

• **1916** Major cities estimate that of the 1000–1750 lb of waste generated by each person per year, 80% is coal and wood ash.

• **1917** Shortages of raw materials during the First World War prompt the US Federal Government to start the Waste Reclamation Service, part of the War Industries Board. Its motto is 'Don't waste waste—save it.' Every article of waste is considered valuable for industry.

• **1935** General Electric begins producing and marketing a garbage disposal unit.

• **1941** The US enters the Second World War. Rationing of such materials as wood and metal forces an increased reliance on synthetic materials such as plastics. Low-density polyethylene film, developed during wartime, replaces cellophane as the favourite food wrap by 1960.

- **1942** Methods and materials for wartime shipment of food make the Second World War the 'great divide' in the packaging and storage industry.
- **1948** The American Public Health Association predicts that the garbage disposal will cause the garbage can to 'ultimately follow the privy' and become an 'anachronism'.
- **1950** The growth of convenience foods (frozen, canned, dried, boxed, *etc*.) increases the amounts and changes the types of packaging thrown away.
- **1965** The US Federal Government realizes that garbage has become a major problem and enacts the Solid Waste Disposal Act. This calls for the nation to find better ways of dealing with trash.
- **1969** Seattle, Washington institutes a new fee structure for garbage pick-up. Residents pay a base rate for one to four cans and an additional fee for each additional bundle or can.
- **1970** The Federal Clean Air Act is enacted. New regulations lead to incineration shut-downs.
- **1972** According to William Ruckelshaus, head of EPA, solid waste management is a 'fundamental ecological issue. It illustrates, perhaps more clearly than any other environmental problem, that we must change many of our traditional attitudes and habits'.
- **1976** The Resource Conservation and Recovery Act is passed, which requires all dumps to be replaced with 'sanitary landfills'. The enforcement of this act will increase the cost of landfill disposal, and that will make resource-conserving options such as recycling more appealing.
- **1986** The city of San Francisco meets its goal of recycling 25% of its commercial and residential waste.
- **1991** Our economy is such that we cannot 'afford' to take care of things: labor is expensive, time is expensive, money is expensive, but materials—the stuff of creation—are so cheap that we cannot afford to take care of them.'—Wendell Berry.
- **2000** Cities in California are required to recycle 50% of their waste.

Source: adapted from http://www.bfi-salinas.com/kids_trash_timeline-printer.cfm

2.4 FOOD WASTE ALONG THE FOOD SUPPLY CHAIN

Food waste is the outcome of many drivers: the market economy, resource limitations and climate, legislation and cultural differences being just a few. It originates along the entire food supply chain, which is a global and complex network of participants, ranging from the individual farmer/manufacturer/producer through to large multinationals. It consists of a series of physical and decision-making activities connected by material and information flows and associated flows of money and property rights that cross organizational boundaries, but depend on the logistic flows—transporters, warehouses, retailers, service organizations and consumers themselves.[21] It is a structured, measured set of activities designed to produce a specified output for a particular customer or market.[22]

Looking at the types of products, Vorst[23] classifies food supply chains into those involved in growing fresh agricultural products and those manufacturing processed foods. The former includes growers, auctions, wholesalers, importers and exporters, retailers and specialty shops in which the handling, storing, packing, transportation and especially trading of these goods are fulfilled. The second type of chain generally uses inputs from the first type to produce consumer goods with higher added value (Figure 2.2).

Attempts have been made over several decades to quantify global food waste. Such assessments are reliant on limited data sets collected across the food supply chain at different times and extrapolated to the larger picture. The most often quoted estimate is that 'as much as half of all food grown is lost or wasted before and after it reaches the consumer',[24] although there is no consensus on the proportion of global food production that is currently lost (Tables 2.3 and 2.4).

In particular, food is wasted in the following ways:

- **Food production** wastes pre-harvest food through natural disasters, diseases, or pests; harvested food by inefficient collection of edible crops or livestock; and post-harvest food in storage or contamination losses.
- **Food processing** wastes food in spillage, spoilage, discarding substandard edible materials, or removing edible food parts in inefficient processing.

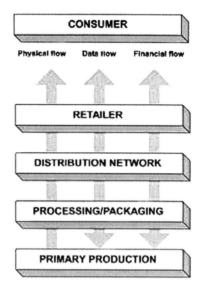

Figure 2.2   Food waste scheme. Source: authors' elaboration.

- **Food distribution** wastes food by offering more food than consumers will purchase and then discarding unsold products.
- **Food acquisition** wastes food when consumers purchase more food than they use.
- **Food preparation** wastes food by removing edible parts of foodstuffs, spilling or contaminating foods, and rendering foods inedible through improper handling and overcooking.
- **Food consumption** wastes food by taking larger portions than can be eaten or by spilling food.

In the literature, post-harvest food waste is likely to be referred to as 'food losses' and 'spoilage'. Food loss refers to a decrease in food quantity or quality that makes it unfit for human consumption.[25] At later stages of the food supply chain, the term 'food waste' is used and generally relates to behavioural issues. Food losses/spoilage, conversely, relate to systems that require investment in infrastructure.

Similarly, both 'food supply chain' and 'post-harvest systems' are used to mean the same thing in the literature, with 'post-harvest loss' also often used when describing agricultural systems and the onward supply of produce to markets.

Table 2.3 Generic food supply chain and examples of food waste

Stage	Examples of food waste/loss characteristics
1 Harvesting—handling at harvest	Edible crops left in field, ploughed into soil, eaten by birds, rodents, timing of harvest not optimal: loss in food quality Crop damaged during harvesting/poor harvesting technique Out-grades at farm to improve quality of produce
2 Threshing	Loss through poor technique
3 Drying—transport and distribution	Poor transport infrastructure, loss owing to spoiling/bruising
4 Storage	Pests, disease, spillage, contamination, natural drying out of food

Processing

Stage	Examples of food waste/loss characteristics
5 Primary processing—cleaning, classification, de-hulling, pounding, grinding, packaging, soaking, winnowing, drying, sieving, milling	Process losses Contamination in process causing loss of quality
6 Secondary processing—mixing, cooking, frying moulding, cutting, extrusion	Process losses Contamination in process causing loss of quality
7 Product evaluation—quality control: standard recipes	Product discarded/out-grades in supply chain
8 Packaging—weighing, labelling, sealing	Inappropriate packaging damages produce Grain spillage from sacks Attack by rodents
9 Marketing—publicity, selling, distribution	Damage during transport: spoilage Poor handling in wet market Losses caused by lack of cooling/cold storage
10 Post-consumer—recipes elaboration: traditional dishes, new dishes product evaluation, consumer education, discards	Plate scrapings Poor storage/stock management in homes: discarded before serving Poor food preparation technique: edible food discarded with inedible Food discarded in packaging: confusion over 'best before' and 'use by' dates
11 End of life—disposal of food waste/loss at different stages of supply chain	Food waste discarded may be separately treated, fed to livestock/poultry, mixed with other wastes and landfilled

Source: Authors' elaboration.

Table 2.4   Estimated total waste arising, by type, from the UK food and drink supply chain, by stage, and household, per year (tonnes)

Supply chain stage	Food	Packaging	Other	Total
Manufacturing	2 591 000	406 000	2 019 000	5 016 000
Distribution	4000	85 000	9000	98 000
Retail	362 000	1 046 000	56 000	1 464 000
Household	8 300 000	3 600 000	20 566 000	32 466 000
Total	*11 257 000*	*5 137 000*	*22 650 000*	*39 044 000*

Source: www.wrap.org.uk

2.4.1   Food Waste at Post-Harvest Stage

Research into the food industry of the US, whose food supply is the most diverse and abundant of any country in the world, found food waste occurring at the very beginning of food production. From the time of planting, crops can be subjected to pest infestations and severe weather, which cause losses before harvest. Since natural forces (*e.g.* temperature and precipitation) remain the primary drivers of crop growth, losses from these can be experienced by all forms of outdoor agriculture. The use of machinery in harvesting can cause waste, as harvesters may be unable to discern between ripe and immature crops, or collect only part of a crop.

Economic factors, such as regulations and standards for quality and appearance, also cause food waste; farmers often harvest selectively, preferring to leave crops not up to standard in the field (where they can be used as fertilizer or animal feed), since they would otherwise be discarded later.

Food waste continues in the post-harvest stage, but the amounts of post-harvest loss involved are relatively unknown and difficult to estimate. Regardless, the variety of factors that contribute to food waste, both biological/environmental and socio-economic, would limit the usefulness and reliability of general figures.

In storage, considerable quantitative losses can be attributed to pests and microorganisms. This is a particular problem for countries that experience a combination of heat (around 30 °C) and ambient humidity (70–90%), as such conditions encourage the reproduction of insect pests and microorganisms. Losses in the nutritional value, calorific value and edibility of crops, by extremes of temperature, humidity or the action of microorganisms, also account for food waste. These 'qualitative' losses are more difficult to assess than quantitative ones. Further losses are generated in the handling of food and by shrinkage in weight or volume.

Post-harvest losses are considered along a technological/economic gradient: 'developing', 'intermediate' and 'industrialized' food supply chains. Table 2.5 provides an overview of the development of post-harvest infrastructure along this gradient.

In developing countries, most of the rural poor rely on short food supply chains with limited post-harvest infrastructure and technologies. Farming is mostly small scale with varying degrees of involvement in local markets and a rapidly diminishing proportion of subsistence farmers who neither buy nor sell food staples.[26] Interventions within these systems focus on training and upgrading technical capacity to reduce losses, increase efficiency and reduce labour intensity of the technologies employed. However, attempts to reduce post-harvest losses must take account of cultural implications. In years with food surpluses, the prices received for goods will be low. One option is to store surplus for lean years, but there may be no suitable storage facilities. To rectify this, investment and engineering skills are needed. There are many instances of relatively simple technologies providing effective solutions.

Transitional and industrialized post-harvest systems have a closer integration of producers, suppliers, processors, distribution systems and markets ensuring greater economies of scale, competitiveness and efficiency in the food supply chain. Supermarkets are the dominant intermediary between farmers and consumers. Even in poorer transitional economies, supermarkets are the main vehicle for delivering diversified diets for the growing middle classes and the urban poor. This is almost entirely dependent on foreign direct investment, with high growth rates in eastern Europe, Asia and Latin America.[27]

The sequence of transformation follows a different route in each country, particularly in the extent to which retailers bypass existing markets and traditional wholesalers to secure produce of the required standard and volume. Accounts of supermarket expansion in some countries suggest there are instances of successful adaptation to traditional supply chains,[28] particularly in regions that have not been so involved in export-orientated markets. Where central wholesale markets are used to source fresh produce, retailers may be reliant on wholesalers to perform the 'out-grading' that in developed countries is likely to occur

Table 2.5 Characterization of post-harvest infrastructure in relation to stages of economic development

Type of post-harvest infrastructure	Technological development	Level of development	Supply chain characteristics	Type of growers	Markets and quality
Developing traditional systems	Simple technologies, labour-intensive, traditional storage systems and harvesting techniques	Low-income countries	Poor integration with local markets, many intermediaries supplying urban markets	Smallholders, including subsistence farmers	Local markets: mostly meeting household/village food requirements; limited access to international markets
Intermediate systems—'transitional'	Packing houses, refrigeration and storage facilities systems alongside elements of traditional systems	Low- and middle-income countries	Requires closer integration of growers, suppliers, processors and distribution systems	Small-scale farmers who often have access to limited post-harvest-specific infrastructure	Produce of variable quality, target both local (including supermarkets) and, increasingly, export markets in a number of countries
Developed industrialized systems	Access to relatively sophisticated technologies, e.g. packing-house equipment and cold chains; losses still occur; harvesting highly mechanized, e.g. wheat	Middle- and high-income countries	Use of highly integrated systems between growers and supply chain; more seasonal produce imported; more secondary processing of foo		

Source: Authors' elaboration.

on-farm or at front-end packing operations. In countries with traditional two-tier produce markets (higher-quality export and lower-quality domestic markets), local supermarkets have created a third market for intermediate to high-quality products. At the same time, retailers provide upward pressure to improve product quality and food safety in the domestic market.

Development of more industrialized food supply chains can also result in growth in the food processing sector. In some of the BRIC countries (Brazil, Russia, India, China), public sector investment is being considered to accelerate this process. In India, the government is discussing an 'evergreen revolution', which will involve the build-up of food processing units.[29] While this is a sensitive issue because of concerns about the industrialized sector taking control over small farmers, the improved infrastructure has helped farmers branch out into new foods, diversifying their incomes.

In medium–high income countries it is often argued that better resource efficiency and less waste are achieved through central processing of food and industrialized food supply chains. Although more food wastage occurs at the factory, logic suggests less waste overall is generated as there is less 'scratch-cooking' at home. However, research on post-consumer food waste suggests that this is not the case, as consumers still waste significant quantities of food, thus potentially negating the benefits of centralized food processing.

Discarded fish from marine fisheries represents the single largest proportion lost of any food source produced or harvested from the wild. The proportion is particularly high for shrimp bottom-trawl fisheries. Mortality among discarded fish is not adequately known, but has, for some species, been estimated to be as high as 70–80%, perhaps higher.[30] Discarded fish alone amounts to as much as 30 Mt per year, compared to total landings of 100–130 Mt.

Feed for aquaculture is a major bottleneck, as there are limitations to the available oil and fish for aquaculture feed.[31] A collapse in marine ecosystems would therefore have a direct impact on the prices of aquaculture products and on its scale of production. There is no indication that marine fisheries today can sustain the 23% increase in landings required for the 56% growth in aquaculture production required to maintain *per capita* fish consumption at current levels to 2050. However, if sustainable, the amount of fish currently discarded at sea could alone sustain more than a 50% increase in aquaculture production. Many of these species could also be used directly for human consumption.

Fish post-harvest losses are generally high at the small-scale level. Recent work in Africa by FAO has shown that regardless of the type of fisheries (single or multispecies), physical post-harvest losses (*i.e.* fish lost for human consumption) are commonly very low, typically around 5%.[32] Downgrading of fish because of spoilage is considerable, however, perhaps as high as 10% or more. Hence, the total amount of fish lost through discards, post-harvest loss and spoilage may be around 40% of landings.

The potential to use unexploited food waste as alternative sources of feed is also considerable for agricultural products. Food losses in the field (between planting and harvesting) could be as high as 20–40% of the potential harvest in developing countries, due to pests and pathogens.[33]

Post-harvest losses vary greatly among commodities and production areas and seasons. In the US, the losses of fresh fruits and vegetables have been estimated to range from 2% to 23%, depending on the commodity, with an overall average of about 12% losses between production and consumption sites.[34] The US total retail, foodservice, and consumer food losses in 1995 were estimated to be 23% of fruits and 25% of vegetables. In addition, losses could amount to 25–50% of the total economic value because of reduced quality. Others estimate that up to 50% of the vegetables and fruits grown end as waste.[35]

Finally, substantial losses and wastage occur during retail and consumption due to product deterioration as well as to discarding of excess perishable products and unconsumed food. While the estimates vary among sources, it is clear that food waste represents a major potential, especially for use as animal feed, which, in turn, could release the use of cereals in animal feed for human consumption.

In 2007, US$148 billion was invested in the renewable energy market, up 60% from the previous year.[36] Recovering energy from agricultural wastes is becoming increasingly feasible at the industrial production level; investments in technology, enhancement of existing systems and innovation in new waste management systems are called for to support this expanding green economy.

2.4.2 Food Waste in the Manufacturing Sector

Based on EUROSTAT data, food waste in the manufacturing sector represents 76 kg *per capita* in the EU.[37] *Per capita* ratios were also calculated at national level, but results yielded by this exercise are very heterogeneous, ranging from 393 kg *per capita* in the Netherlands to 7 kg *per capita* in Greece. This high heterogeneity could be consistent with the geographic distribution of the EU food industry, which is highly concentrated in certain countries, such as the Netherlands, and less in others, such as Greece.

Food waste is largely unavoidable (inedible) at the manufacturing level, particularly for meat products, involving principally bones, carcasses, and organs that are not commonly eaten. Technical malfunctions also play a role, including overproduction, inconsistency of manufacturing processes leading to misshapen products or product damage, packaging problems leading to food spoilage, and irregular sized products trimmed to fit or discarded entirely (Table 2.6).

In the absence of any European requirement, no systematic data on food processing waste are provided by Member States.

2.4.3 Food Waste in the Retail Sector

Large retailer outlets—supermarkets and hypermarkets—are central in the food supply chain. The supermarket was an idea which generated in the US in the 1920s and spread to Europe after the Second World War. In 1970s stores became bigger and in the 1980s they took control of functions that had traditionally been performed by other manufacturers, including distribution, packaging, advertising and product design.[38] The supermarkets brought about a cultural change, giving previously unimaginable increase in choice, range, convenience, aseasonality (year-round sourcing), and redefinition of quality (appearance).[39]

Our lifestyles, diets and health are influenced by the rise of the ready meal, the pre-packed sandwich and the exotic fruits that the chill chain makes possible. Distant economies depend on this too. Debates about food distribution are never only about distance, and they are rarely just about the environment.

The situation among retailers is changing particularly rapidly. In the period 1993–9, the aggregate concentration of the top 10 grocery retailers in the EU grew by 24.9% whereas the market share of

Table 2.6 Food waste generation in manufacturing sector, total and percentage wasted

	Food production in tonnes (EUROSTAT 2006)	Food Manufacturing Sector tonnes (EUROSTAT 2006)	Manufacturing sector FW tonnes (WRAP data)	Population (EUROSTAT 2006)	FW per capita (EUROSTAT 2006)	% of food wasted with (EUROSTAT) data	% of food wasted with (WRAP) data
EU-27	766 179 686	37 307 575		493 194 250	76	5	
Austria	9 914 359	570 544		8 254 298	69	6	
Belgium	27 470 839	2 311 847		10 511 382	220	8	
Bulgaria	4 849 152	358 687		7 718 750	46	7	
Cyprus	0	186 917		766 414	244		
Czech Republic	13 034 071	361 813		10 251 079	35	3	
Denmark	9 103 122	101 646		5 427 459	19	1	
Estonia	1 143 852	237 257		1 344 684	176	21	
Finland	9 845 332	590 442		5 255 580	112	6	
France	106 199 337	626 000		63 229 443	10	1	
Germany	136 078 334	1 848 881		82 437 995	22	1	
Greece	6 170 557	73 081		11 125 179	7	1	
Hungary	11 702 284	1 157 419		10 076 581	115	10	
Ireland	5 382 309	465 945		4 209 019	111	9	
Italy	97 088 841	5 662 838		58 751 711	96	6	
Latvia	1 606 037	125 635		2 294 590	55	8	
Lituania	4 020 685	222 205		3 403 284	65	6	
Luxemburg	0	2665		469 086	6		
Malta	0	271		405 006	1		
Netherlands	50 834 267	6 412 330		16 334 210	393	13	
Poland	47 233 940	6 566 060		38 157 055	172	14	
Portugal	12 496 826	632 395		10 569 592	60	5	
Romania	10 845 823	487 751		21 610 213	23	4	
Slovakia	3 841 080	347 773		5 389 180	65	9	
Slovenia	1 176 515	42 072		2 003 358	21	4	
Spain	101 939 483	2 170 910		43 758 250	50	2	
Sweden	5 197 871	601 327		9 047 752	66	12	
United Kingdom	87 004 770	5 142 864	2 591 600	60 393 100	85	6	3

Source: 2006 EUROSTAT data

the bottom 10 companies in the EU top 50 declined by 72.2%.[40] The larger are getting larger and the small are being squeezed.[41] In Europe, retailers are now concentrating regionally, perhaps due to the fact that home markets were already concentrated (Table 2.7).

In western Europe, the grocery retail value grew by about 20% between 1999 and 2006: in 1999 it was €8 904 964 million, and by 2006 €1 064 288 million. In the same period, the percentage of retail value taken by the big retailers remained very large, edging up from 68% to 72%. In eastern Europe, by contrast, the grocery market retail value doubled between 1999 and 2006:[42] in 1999 it was worth €87 332 million, and by 2006 it was €179 863 million. In the same period, the percentage of the retail value taken by big retailers also nearly doubled, growing from 23.6% to 44.9%. The early 21st century has seen a shift to supermarkets and hypermarkets and a decline in traditional outlets such as independent food stores, kiosks, and street markets, as eastern Europe—like other regions of the world—follows the supermarket route.[43]

These trends are likely to continue. The Institute of Grocery Distribution,[44] a food sector research institute, predicts that, based on historical growth rates in European turnover for the last 5 years, the top 10 retailers have increased market share from 37 to 60% by 2010. Their combined European grocery turnover has grown from €337.1 billion in 2000 to €461.7 billion by 2005 and €669.7 billion by 2010.[45]

Carrefour, Aldi, Tesco are the emerging European giants. The UK's Tesco, for instance, is now structured into three divisions: UK and Ireland, central Europe and the Far East. In the case of national market concentration by grocery retailers, some earlier figures (from 2002) from consultants Cap-Gemini show that the degree of concentration varies from country to country in western Europe.[46] The highest concentration is found in Sweden and the lowest in Greece and Italy. Much current market concentration has occurred not by slow gains due to superiority of product or consumer appeal, but by buy-outs. Mergers and acquisitions have been rife from the 1980s on both sides of the Atlantic, as already large companies snapped up competitors. The results have changed both the architecture of the food supply chain and its public face.

In the UK alone at least 3 Mt of produce is thrown away by the retail sector,[47] including supermarkets, and food manufacturers.

Table 2.7 Share of food sales for retailers in selected international markets, 2002

Retail outlets	United States	Western Europe	Latin America	Japan	Indonesia	Africa and Middle East	World
				Percent sales			
Supermarkets/hyprmarkets	62.1	55.9	47.7	58.0	29.2	36.5	52.4
Independent food stores	10.0	10.0	33.0	11.3	51.1	27.1	17.8
Convenience stores	7.5	3.8	3.1	18.3	4.8	10.0	7.5
Standard convenience stores	5.7	2.5	1.8	18.2	4.8	9.5	6.4
Petrol/gas/service stations	1.8	1.2	1.3	0.1	0.0	0.5	1.1
Confectionery specialists	0.5	2.0	1.7	0.3	0.1	1.3	1.2
Internet sales	0.2	0.1	0.1	0.4	0.0	0.0	0.2
Chemists/drugstores	0.2	0.3	0.2	0.4	0.2	0.3	0.3
Home delivery	0.4	0.2	0.0	0.0	0.0	0.0	0.1
Discounters	7.4	10.3	0.2	2.2	2.7	6.2	5.7
Other	12.0	17.5	14.0	9.0	11.9	18.6	14.9
Total	100	100	100	100	100	100	100

Source: Euromonitor, 2004.

Areas where food waste is generated include those common to both the wholesale/retail sector and at the manufacturing/processing level. Excess stock due to 'take-back' systems and last-minute order cancellations, such as contractual obligations for suppliers to accept the return of products with 75% residual shelf life from retailers who have not yet sold them, can result in the discard of safe and edible food products on a large scale. Inaccurate ordering and forecasting of demand also affects the wholesale/retail sector.

Other reasons for food waste generation include the following:

Forecasting Difficulties and Poor Ordering. Poor forecasting is one of the most common issues identified as a cause of waste. Estimating the demand for a product is a complex and inherently inaccurate task which can be affected by many factors, such as weather, seasonality, marketing campaigns, product launches, promotions and special occasions such as Christmas and Easter.

A variety of forecasting practices exist in the industry, with some companies using a scientific approach while others use more informal methods. Improving forecasting practices and using up-to-date data mining models can reduce forecast error; however, it has to be recognized that uncertainty will continue to exist and that forecast error cannot be eliminated.

Performance Measurement and Management. The emphasis in the industry appears to be on cost, efficiency and availability. Although waste has an impact on all of these factors, it is not usually a key performance measure and it can be sacrificed at the expense of other performance indicators. For instance, interviews revealed that most mainstream retailers have policies of only accepting product with a high proportion of shelf life remaining (usually over 70%).

Cold Chain Management. Cold chains can help in keeping certain products fresh and avoiding spoilage. On the other hand, mismanagement of the cold chain, caused by equipment failure or poor processes, will inevitably generate waste. Failure to maintain the cold chain (which can be mitigated by the development and investment in new and more reliable technologies) can have a severe impact on waste, but these situations are relatively rare. In Spain, where temperatures tend to be higher than in the UK,[48] problems with the cold chain appear more frequently, particularly with vehicles delivering the products to supermarkets, but not so much with the products stored in warehouses, where more powerful equipment is traditionally used. More investment is needed at this point to reduce this risk.

Training.   In some cases, employees may not follow procedures for stacking, shelving and stock rotation, all of which can lead to waste. This issue appears to be more prevalent during the Christmas period when temporary labour is hired to cope with high demand, and in some specific roles (such as butchers or fishmongers) where good performance can save quality products from spoilage.

Quality Management.   Quality issues can lead to rejections and even product recalls. Rejections in particular appear to be prevalent in milk processing plants (where chemical analysis and sterility of products is very strict and quality standards are very high) as well as in the fruit and vegetable sector, where product quality can be variable (particularly at the beginning and end of seasons), with the producers trying to send to market all the harvest with its variable quality.

   While quality issues can lead to waste, the loss of product quality appears to be more important to the companies than the waste created. Product recalls are relatively rare events. However, when they occur they are likely to generate large amounts of waste, particularly for products with a long shelf-life, since companies are likely to have more stock in the pipeline.

Waste Management Responsibilities.   While some companies have very clear roles and responsibilities for managing waste, others do not have a specific role within the company focusing on waste. This usually means that waste is not measured and managed systematically and this situation is likely to lead to increased waste. An example of this is the case of a vendor managed inventory (VMI) system in operation in Spain,[49] where a higher risk of generating waste was detected because retailers are not responsible for managing stock in their shops and therefore less attention is paid by the people closer to the physical product.

   Stock transportation can lead to both packaging and storage problems. Poor packaging performance resulting in damage to food products will lead to the discard of the product. As noted earlier, damage to the product's primary or secondary packaging also often means the product will be discarded, while the food itself is unharmed. It is generally expected, however, that packaging materials have been optimized to minimize waste and hence waste is expected to be limited here.

Marketing Standards. Aesthetic issues or packaging defects cause some products to be rejected, although neither food quality or safety is affected.

Marketing Strategies. Strategies such as two-for-one deals often promote food nearing the end of its edible life, addressing overstocking problems. However, this may shift some of the food waste from retail level to household level, where there is insufficient time to consume the product safely.

Recent innovations have been launched in order to precisely measure the freshness of food items as they pass through the supply chain from factory to consumer and could lead to a significant reduction in the amount of waste produced.

Processors have little control over the temperatures their goods are exposed to throughout the value chain. Consequently, they often mark their products with a shorter shelf life as a precautionary measure, which can mean a lot of edible food being thrown away. Norwegian food retailers discard over 50 000 t of food annually.[50] A new intelligent technology called TimeTemp,[51] launched by a Norwegian company, gives a running countdown of a product's shelf life by analysing time and temperature data, and could replace traditional 'use by' dates on food labels. The innovative device, which is a small self-adhesive label attached to food products, contains a range of non-toxic chemicals which react and change colour according to time and temperature. The chemical reaction is activated at the food producer's packaging line and follows each item from production to consumer. The reaction shows the time left before expiration of that product in accordance with the actual degradation of the food item, depicted in an easy-to-read graphical format.

Intelligent packaging technology is applicable for all products where quality and lifespan depend on time and temperature variables during storage, as well as items where quality depends on maturity and ageing. Items such as meat, poultry, dairy and even bakery products would potentially benefit from using the technology.

2.4.4   Food Waste at the Consumer Level

Household food waste can be defined as food and drinks that are consumed within the home; this excludes significant quantities of food and drink eaten 'on the go', in the workplace or in catering establishments. Wherever possible, the distinction is made between three classifications of household food waste: 'avoidable', 'possibly avoidable' and 'unavoidable'.

Garbology,[52] the study of human waste behaviours, identifies food waste as a significant portion of the total human waste stream. Food waste comprises about 10% of the total municipal solid waste streams in post-industrial nations and higher percentages in societies lacking mechanized refrigeration and durable packaging.[53]

'Out of sight, out of mind' describes the common human-food consumed/trash relationship. From 1987 to 1995, the Garbage Project, a University of Arizona nationwide study,[54] found that self-reports of eaten food did not match the garbage generated. Like nutrition assessments, discrepancies occurred with recall accuracy, selective reporting and withholding of information. Perceived 'good' behaviours were over-reported while 'bad' behaviours were under-reported. Household matriarchs over-report all food-related behaviours by 10–30%.[55]

Garbage data can reveal information about food behaviours of groups, time periods, and households, as well as nutrition adequacy and cost-effectiveness. To estimate the proportion of food wastes at the household level, methodologies such as enquiries or waste sorting analyses can be used. The results of enquiries give qualitative information like kind and frequency of wasted food and reasons for wasting food on the basis of self-reported behaviour of the respondents. Furthermore, information about interrelated conditions (*e.g.* level of employment, age of household members) and behaviour (*e.g.* buying, cooking and diet habits) can be gathered.

The method of waste sorting analysis of residual waste is used to find information about the quantitative composition of residual waste where these data are important for monitoring and planning of waste management systems. This method only gathers data about the items disposed of into the residual waste bin; other disposal paths such as a garbage disposal unit in the kitchen sink, pet food, home composting and source-separated collection are not considered.

In the UK, detailed research findings[56] highlight that the two principal reasons why avoidable food waste occurs are: 'food is not used in time' and 'too much food is cooked, prepared or served':

- **Cooked, prepared or served too much.** In the majority of cases, this is because too much food was 'processed' in the home, but also covers cases where food was damaged during processing (*e.g.* burning food). This category could be referred to as 'leftovers'.
- **Not used in time.** This covers food and drink wasted because it passed a date label (*e.g.* a 'use by' or 'best before' date), has gone mouldy or looked, smelt or tasted bad.

Causes for waste include:

- **Lack of awareness** of (1) the quantity of food waste generated individually, (2) the environmental problem that food waste presents, and (3) the financial benefits of using purchased food more efficiently.
- **Lack of knowledge** on how to use food efficiently, *e.g.* making the most of leftovers, cooking with available ingredients.
- **Attitudes:** food undervalued by consumers, lack of necessity to use it efficiently.
- **Preferences:** many (often nutritious) parts of food are discarded due to personal taste, *e.g.* apple skins, potato skins, bread crusts.
- **Planning issues:** 'buying too much' and 'lack of shopping planning' frequently cited as causes of household food waste.
- **Labelling issues:** misinterpretation or confusion over date labels is widely recognized as contributing to household food waste generation, leading to the discard of still edible food.
- **Storage:** inappropriate storage conditions leads to food waste throughout the supply chain and is no less important in the household. Lack of consistency in food storage labels can contribute to premature food spoilage, as can the absence of

storage guidance and lack of consumer attention to labels where provided. Storage conditions will also vary based on climate and household temperature. WRAP reports that over 2 Mt of food is not being stored correctly in the UK, multiplying food wastage and presenting potential safety concerns. Optimal storage conditions, by contrast, can significantly extend the edible life of products, often beyond expiry dates. Airtight containers, for example, easily maintain the quality of dry foods such as fruits, nuts, rice, pasta, beans and grains over long periods.

- **Packaging issues:** packaging methods and materials can impact the longevity of food products. The lifetimes of products with a high water content, cucumbers for example, can be extended fivefold through plastic film wrapping, as it reduces water loss. Packaging also performs a protective function for fragile goods. The trade-off between food and packaging waste must then be considered, based on the environmental impacts of the two waste streams, though this again will be highly product specific. In some instances, lightweight packaging can significantly extend the shelf life of fresh produce; in other cases the benefit can be marginal.

- **Portion sizes:** includes issues such as 'making too much food' hence leading to uneaten leftovers as well as purchasing the correct portions of food; individually sized portions can minimize food waste but often create additional packaging waste. Bulk packaging minimizes the ratio of packaging to food product delivered to the consumer, though the quantity may be greater than the consumer can use while the product is fresh. Individually sized portions can minimize food waste, but create extra waste in another waste stream (plastics, glass *etc.*). Better storage knowledge, freezing and preserving information, and storage equipment in the household can help purchases last longer and minimize reliance on smaller portions.

- **Socio-economic factors:** single-person households and young people generate more food waste.[57]

The labelling issue in particular is a very serious one. Misinterpretation or confusion over date labels is widely recognized for its contribution to household food waste. In many Member States there is a lack of consistency in the terms employed ('best before', 'use by', 'sell by', 'display until'). There is a tendency among consumers to treat all terms equally, and in some cases to leave a safety margin before the stamped date. Using 'best before' dates for products that show visible signs of decay may be unnecessary, causing consumers to discard something that does not pose a safety risk. Consumers might be better left to judge the quality and safety of such products for themselves—bread or potatoes, for example. By contrast, the use of 'best before' dates on products that are liable to pose microbiological risks after a certain date, eggs or yoghurt for example, is also a concern. In this scenario, consumers may consider the date as a quality indicator, when in fact the product may have become dangerous.

At the point where consumers decide whether to eat or discard a food product in the household, sensory judgements on the quality and safety of the food will interplay with an assessment of the date label on the product. A lack of clarity and consistency in date labels thus results in a greater proportion of discarded food that was in fact still edible. Figure 2.3 shows the interaction of criteria used in assessing product edibility.

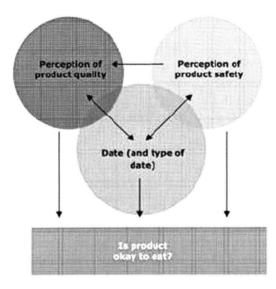

Figure 2.3 Deciding if a product is edible. Source: Research into Consumer Behaviour in Relation to Food Dates and Portion Sizes (WRAP 2008).

There is often a large variation in the wastage rates for different food types: WRAP found that 7% of milk purchases are wasted, 36% of bakery products and over 50% of lettuce/leafy salads (by weight),[58] while Jones[59] found similar variations in the average wastage rates for different food types.

The following factors may help to explain variation in quantities of household food waste generated.

- **Household size and composition.** Studies from the UK[60] and the USA[61] show that food wastage was significantly influenced by the composition of the family, with adults wasting more in absolute terms than children, and larger households wasting less per person than smaller households. Single-person householders tend to throw away more *per capita*, and households with children tend to waste more than households without children, although rates vary with the children's age.
- **Household income.** The majority of studies suggest that there is lower food loss in low-income than in high-income households;[62] other studies[63] found little or no correlation between income and food wastage.
- **Household demographics.** Studies in the UK[64] and Australia[65] suggest that young people waste more than older people, with pensioner households wasting the least (such households normally contain comparatively fewer people).
- **Household culture.** There is some indication that culture partly determines food wastage. For example, Hispanic households in the USA have lower food loss rates[66] (approx. 25% less) than non-Hispanics.

Some studies have measured household food waste as a percentage of total consumed calories, others as a percentage of the total weight of consumed food or of the consumed food items. Some studies have sought to estimate the environmental impact of food waste, including the embodied greenhouse gas emissions or water.[67]

Methodologies for post-consumer waste analysis vary, from small numbers of households weighing food waste or using kitchen diaries to waste compositional and behavioural studies involving thousands of households. Others have used contemporary archaeological excavations of landfill sites to determine historical levels of food waste:[68] estimated household food waste indirectly from loss coefficients based upon existing research;[69] or estimated wastage using statistical models relating population metabolism and body weight.[70]

Increased consumer choice and a decrease in the proportion of disposable income spent on food have tended to increase wasteful behaviour.

2.4.5 Food Service Businesses and Institutions

The food service industry estimates that 4–10% of food purchases become waste[71] before ever reaching a guest. Food waste occurs at different levels and for different reasons:

- According to a 2005 study at Cornell University,[72] consumers eat 92% of the food they serve themselves, but they throw away much more food when portion sizes are imposed.
- Where a self-service option is not viable, a choice of portion size may reduce food waste generation by recognizing that individuals have different portion needs.
- Where food is served *via* a buffet, customers often expect that nothing will run out, particularly in the luxury market, causing businesses to prepare and cook substantially more than will be consumed.
- A final logistical issue in restaurants is cooking according to the 'just in time' principle. Where food is overcooked or not cooked at the same time as the rest of the table's dishes, it is commonly discarded and the process is restarted.
- The practice of taking home restaurant leftovers, a practice that would enable substantial reduction of restaurant food waste, is frowned upon in some parts of Europe.

2.5 FOOD WASTE: A GLOBAL ISSUE

In the following sections, facts and figures regarding food waste from selected countries are presented.

2.5.1 Food Waste in Europe

The total amount of food waste in the EU-27 is estimated at 89 Mt (Figure 2.4), *i.e.* 179 kg per head per year.

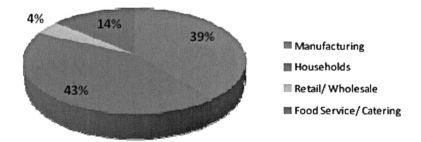

Figure 2.4   Percentage breakdown of EU27 food waste arising by manufacturing, households, retail and food service. Source: EUROSTAT
data, various national sources.

Most food waste from the manufacturing sector is inedible,
and thus could not be avoided, except by transforming it into
by-products. Unfortunately, the existing data collection systems
are not reliable enough to ensure that by-products are not included
in waste figures, so the figure of 39% must be considered with
caution. Manufacturing food waste was estimated at almost 35 Mt
per year in the EU-27 (70 kg per head), although a lack of clarity
over the definition of food waste (particularly as distinct from
by-products) among Member States makes this estimate fragile
(Table 2.8).

The figure for households seems more reliable. EU households
produce the largest proportion of food waste among the four
sectors considered, at 43% of the total or about 38 Mt, an average
of about 76 kg per head per year. Food which ends up as being
discarded by households represents 25% of food purchased (by
weight), according to studies completed by WRAP.[74]

The remaining food waste generated in the EU is estimated to be
split as follows:

- **Wholesale/retail sector:** close to 8 kg *per capita* (with an
  important discrepancy between Member States) representing
  an overall 3 772 685 kg for the EU-27.
- **Food service sector:** more than 28 kg *per capita* in the EU-15
  (due to a higher trend of food waste in the restaurant and
  catering sector) and 12 kg *per capita* in the EU-12, with an
  average of 25 kg *per capita* for EU-27 and an overall 12 263 210
  kg for the EU-27.

Table 2.8 Total food waste generation in EU Member States

	Manufacturing	Households	Retail/ Wholesale	Food Service/ Catering	Total
EU27	34 755 711	37 701 761	3 772 685	12 263 210	88 493 367
Austria	570 544	784 570	267 000	103 500	1 725 614
Belgium	2 311 847	934 760	93 417	287 147	3 627 171
Bulgaria	358 687	288 315	68 598	92 472	808 072
Cyprus	186 917	47 819	6811	9182	250 730
Czech Republic	361 813	254 124	91 104	122 810	829 851
Denmark	101 646	494 914	45 676	148 266	790 502
Estonia	237 257	82 236	11 951	24 564	356 008
Finland	590 442	214 796	46 708	143 570	995 515
France	626 000	6 322 944	561 935	1 080 000	8 590 879
Germany	1 848 881	7 676 471	72 000	2 000 000	11 597 352
Greece	73 081	412 758	98 872	303 914	888 625
Hungary	1 157 419	394 952	89 553	120 720	1 762 643
Ireland	465 945	292 326	37 407	114 981	910 658
Italy	5 662 838	2 706 793	522 140	1 604 960	10 496 732
Latvia	125 635	78 983	20 393	27 490	252 500
Lithuania	222 205	111 160	30 246	40 772	404 383
Luxembourg	2665	62 538	4169	12 814	82 186
Malta	271	22 115	3599	4852	30 838
Netherlands	6 412 330	1 837 599	145 166	446 213	8 841 307
Poland	6 566 060	2 049 844	339 111	457 130	9 412 144
Portugal	632 395	385 063	93 934	288 737	1 400 130
Romania	487 751	696 794	192 055	258 895	1 635 495
Slovakia	347 773	135 854	47 895	64 564	596 086
Slovenia	42 072	72 481	17 804	11 405	143 763
Spain	2 170 910	2 136 551	388 890	1 195 374	5 891 725
Sweden	601 327	905 000	110 253	298 880	1 915 460
United Kingdom	2 591 000	8 300 000	366 000	3 000 000	14 257 000

Source: EUROSTAT data, various national sources and adapted from European Commission [DG ENV—Directorate C] Final Report—Preparatory Study on Food Waste, October 2010.

Methodologies for collecting and calculating the food waste data submitted to EUROSTAT differ between Member States, which are free to choose their own methodology. Limitations in the reliability of EUROSTAT data, due to a lack of clarity on the definition and methodology, may be significant. Implications of this may involve the inclusion of by-products, green waste or tobacco in the data disclosed in some instances. Additionally, data is missing for some sectors in some Member States, and the 'Other

Sectors' category is too broad to give a clear insight into the wholesale/retail and food service sectors. It was not possible to confirm that by-products were not included in some instances in the manufacturing sector data.

Occasionally national studies presented data *per capita* but not total data. In these instances, the *per capita* figure was multiplied by the 2006 population of that Member State, as recorded on EUROSTAT. Calculations were made as follows:

- **Estonia:** The Saastva Esti Instituut (SEI 2008) reports 30% of mixed municipal waste is kitchen waste, and EEIC 2008 states 356 000 t of mixed municipal waste were generated in Estonia in 2008, an average 30% of this comes to 106 800 t. This was subsequently divided between the household and food service sectors.[75]

- **France:** ADEME, the French Environment Agency, reports household food waste in France is at 100 kg per head per year; this has been multiplied by the population of France in 2006, totalling 6 322 944 t.[76]

- **Ireland:** A study by the Clean Technology Centre for the Irish EPA shows that food waste is 16.6% of household municipal waste, which is stated as 1 761 000 t in 2008, resulting in 292 326 t of food waste.[77]

- **The Netherlands:** The Dutch Environment Ministry Food Waste Report presents household food waste as 76–149 kg per head per year. An average of 112.5 kg was thus used to generate the national total. Although among the higher figures, this compares reasonably well to the EUROSTAT figure of 104 kg *per capita*.[78]

- **Sweden:** The Naturvardsverket (2010) study identifies household food waste in Sweden at 100 kg per head per year, and this was multiplied by the Swedish population in 2006, giving a total of 905 000 t.[79]

Food Waste in the UK. The most comprehensive data on food waste is presented by WRAP in the UK. Their 2010 study on supply chain food waste[80] presents up-to-date quantities of food waste in the manufacturing, wholesale/retail and household sectors. Published at the end of March 2010, the quantities presented here have been updated to reflect changes in UK food waste estimates as a result of this study. The current WRAP estimate on the food service sector is unchanged at 3 Mt, although a WRAP study on food waste arising in the hospitality industry is currently under way which may significantly change this figure.

Pre-Second World War studies[81] showed that 1–3% of food was wasted in the home in Britain. The next major study, by the UK Ministry of Agriculture, Fisheries and Food in 1976, investigated the 25% 'crude energy gap' between estimates of embodied energy in domestically grown and imported food (an average of 12.3 MJ (2940 kcal) of energy for each person per day), and the average physiological requirement for energy according to the UK Department of Health and Social Security (9.6–9.8 MJ (2300–2350 kcal) per person per day).[82] The resultant survey of 672 households recorded all the potentially edible food wasted in a week, and found that, when assessed against the expected usage of food in the home, wastage accounted on average for 6.5% of the energy intake in summer and 5.4% in winter.[83]

More recently, WRAP has shown that household food waste has reached unprecedented levels in UK homes,[84] with 8.3 Mt of food and drink wasted each year (with a retail value of £12.2 billion at 2008 prices) and a carbon impact exceeding 20 Mt of CO_2 equivalent emissions. The amount of food wasted per year in UK households is 25% of that purchased (by weight).

Food and drink waste is estimated to be approximately 14 Mt in the UK (Figure 2.5), of which 20% is associated with food processing, distribution and retail. Household food waste makes the largest single contribution, but reliable estimates of other post-consumer wastes (hospitality, institutional sources) have yet to be published. The estimated total waste arising from the food and drink manufacturing and processing sector is 5 Mt per year, where approximately 2.6 Mt is estimated to be food waste; a further 2.2 Mt of by-products are diverted into animal feed.[85]

Every day, Britain throws away 220 000 loaves of bread, 1.6 m bananas, 5500 chickens, 5.1 m potatoes, 660 000 eggs, 1.2 m sausages and 1.3 m pots of yoghurt. Table 2.9 shows that 18% or so of UK household waste is food waste (the amount per household is 216 kg).[86] Waste production surveys have identified that a large

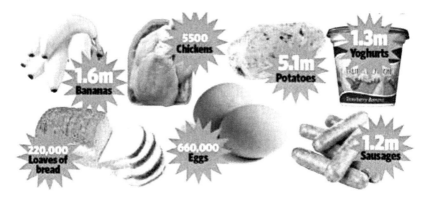

Figure 2.5 Food waste in the UK. Source: http://www.independent.co.uk/life-style/
food-and-drink/news/what-a-waste-britain-throws-away-16310bn-
of-food-every-year-822809.html

Table 2.9 Data and performance estimates for UK in respect of food
waste, 2004/5

	England	Wales	Scotland	N. Ireland	UK
Household waste (000 t)	25 688	1585	2276	919	30 468
Composition of food waste (% hhld)	17.5%	18%	18%	19%	17.6%
Quantity of food waste (000 t)	4495	285	410	184	5375
Average food waste per household per year	—	—	—	—	216 kg

Source: http://www.wrap.org.uk/downloads/Dealing_with_Food_Waste_-_Final_-_2_March_
07.f11ca879.3603.pdf

proportion of food waste originates from the meat, poultry and
beverage sectors. These wastes largely consist of by-products and
unsold prepared food products.

As an indication of the overall resource efficiency of the sector, a
mass balance estimated that nearly 56 Mt of ingredients are used
annually to produce 59 Mt of food products.[87] More mass balances
conducted at food and drink manufacturing sites suggested that
around 16% of raw materials were wasted.[88]

At the retail and distribution stage, the most recent estimate
suggests 366 kt per year.[89] The amounts of waste produced by food

retailers vary between outlet types. Small grocery stores produce proportionately more waste than large supermarkets, as the former tend to be used by consumers for top-up shopping, which makes demand unpredictable.

There is limited information available on the amount of waste generated by the agricultural sector.

Food Waste in Italy. Italian shops and restaurants dispose of a quantity of food which is 88% more than the food the population actually needs. This represents a surplus of 1700 kcal per day: some of these extra calories are consumed by people who eat more than is necessary for their body, but most of them are wasted in the form of leftover food on plates and bags full of still edible products outside the supermarkets. Italy produces a quantity of food which is 3.3 times higher than it requires. In Italy alone, according to *Il Libro nero degli sprechi* 2010 (The Italian Food Waste Report 2010),[90] 240 000 t of food worth over €1 billion remain unsold in back rooms of stores: such food could feed 600 000 people with three meals a day for a year. Food waste in Italy amounts to 3% of GDP.

Every day 4000 t of food, 15% of bread and pasta, 18% meat and 12% of vegetables and fruit end up in landfills or incinerated.[91] According to the Association for the Defence and Orientation of Consumers (ADOC) in Italy, each household throws away €584 a year for food that goes wasted—equivalent to 11% of the yearly expenditure on food (€5400).[92]

Finally, the estimate for the food waste generated by distribution channels is 263 645.22 t, generated by the different stakeholders in the distribution chain.

2.5.2 Food Waste in Asia

India. Some 23 Mt of food cereals, 12 Mt of fruits and 21 Mt of vegetables are lost each year, with a total estimated value of 240 billion rupees.[93] Losses for cereals and oil seeds are lower, about 10–12%, according to the Food Corporation of India. A recent estimate by the Ministry of Food Processing is that agricultural produce worth 580 billion rupees is wasted in India each year.[94]

Australia. Despite the dearth of food waste data in Australia, a submission to a Senate inquiry estimated that food waste comprises 15% of the 20 Mt of waste that goes to landfill each year.[95]

South Korea. A study[96] followed the 2002 South Korean landfill ban on food waste in the municipal waste stream and suggested that food accounted for 26–27% of household waste.[97] Despite an awareness-raising effort in advance of the ban, food waste increased by almost 6% over 4 years after the ban, with increased consumption of fresh fruit and vegetables linked to higher incomes cited as a reason.

Japan. Every year, Japan produces about 400 Mt of industrial waste and about 50 Mt of household and general commercial waste. Of the household and general commercial waste, about 20 Mt consist of food waste.[98] For food waste from the processing and manufacturing' stage, 48% is recycled. Examples of this include feeding pigs fresh bean curd pressings that are by-products of making tofu.

On the other hand, most food waste from the distribution stage is not recycled. Examples include unsold food at supermarkets and lunch boxes and hamburger meat that have passed the expiry date at convenience stores. This is mostly incinerated and sent to landfills.

Under Japan's Basic Law for Establishing a Recycling-Based Society,[99] which came into force in January 2003, the Food Recycling Law took effect in June of the same year. This law's aims are to reduce the amount of food waste generated by food manu-facturers and restaurants, and to promote the re-use of food waste such as by turning it into livestock feed and compost. Pressed into action by this law, an increasing number food manufacturers and restaurants are working to use food waste as compost.

2.5.3 Food Waste in the US

A 1998 study by Kantor *et al.* of food waste in the USA revealed that 25% of food was wasted.[100] Archaeological excavations of US landfills by the University of Arizona[101] also drew attention to food waste in the USA and provided quantitative data on the likely scale.

In the US 30% of all food, worth US$48.3 billion (€32.5 billion), is thrown away each year. It is estimated that about half of the water used to produce this food also goes to waste, since agriculture is the largest human use of water. Losses at the farm level are probably about 15–35%, depending on the industry. The retail sector has comparatively high rates of loss of about 26%, while supermarkets, surprisingly, only lose about 1%. Overall, losses amount to around US$ 90 billion–100 billion a year.[102]

Jones estimated that American households discarded 211 kg of food waste per year, not including food to drain, home composting or feeding to pets. The amount of food loss at the household level was estimated to be 14%,[103] costing a family of four at least $589.76 annually (Jones 2004). The US Environmental Protection Agency estimated that food waste in 2008 accounted for 12.7% (31.79 Mt) of the municipal solid waste stream (Figure 2.6).[104]

On average, each American consumes about 3 lb (1.4 kg) of food each day. If even 5% of the 96 billion lb were recovered, that quantity would represent the equivalent of a day's food for each of 4 million people. Recovery rates of 10% and 25% would provide enough food for the equivalent of 8 million and 20 million people, respectively.

Food is lost at every stage of the US marketing system. However, because of the enormous size and diversity of the American food industry, few studies estimate aggregate marketing losses across the

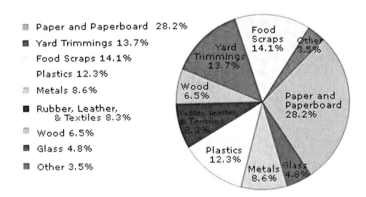

Figure 2.6 Total US Municipal Waste Generation by Material, 2009: 243 Mt (before recycling). Source: http://www.epa.gov/epawaste/conserve/materials/organics/food/fd-basic.htm

entire food sector. Typically, researchers report food losses as a percentage of food servings, household food stocks or retail inventories at specific points in the marketing system, such as fresh fruit and vegetable losses in supermarket produce departments, household plate waste, or preparation and storage losses in food-service operations.

Food has long been relatively cheap, and portions are increasingly huge. With so much news about how fat everyone is getting—66% of adult Americans are overweight or obese, according to a 2003–04 government health survey[105]—there was a compelling argument to be made that it was better to toss the leftover deep-dish pizza than eat it for lunch the next day.

2.5.4 Food Waste in Africa

In many African countries, the post-harvest losses of food cereals are estimated at 25% of the total crop harvested. For some crops such as fruits, vegetables and root crops, which are less hardy than cereals, post-harvest losses can reach 50%.[106] In East Africa and the Near East, economic losses in the dairy sector due to spoilage and waste could average as much as US$90 million/year. In Kenya each year around 95 million litres of milk, worth around US$22.4 million, are lost.[107]

Cumulative losses in Tanzania amount to about 59.5 million litres of milk each year, over 16% of total dairy production during the dry season and 25% in the wet season. In Uganda, approximately 27% of all milk produced is lost, equivalent to US$23 million/year.[108]

2.6 THE ENVIRONMENTAL COST OF FOOD WASTE

In terms of greenhouse gas (GHG) emissions, it is estimated that the overall impact of food waste is 170 Mt CO_2 equivalent per year, *i.e.* 3% of total EU-27 emissions in 2008.[109] In particular, the impact of household food waste is estimated at 78 Mt CO_2, *i.e.* 45% of the total GHG emissions due to food waste. As a great part of this waste is avoidable, from these figures we can conclude that a substantial reduction of food wastage at household level could reduce the total GHG emissions in the EU-27 by nearly 1%. However, once again, it is important to note that these figures have to be taken very cautiously and that the true costs of food waste are generally not perceived. The cost of food waste is reflected in the value of the human effort, capital, energy and raw material that are lost.

Environmentally, food waste leads to wasteful use of chemicals such as fertilizers and pesticides; more fuel used for transportation, and more rotting food, creating more methane—one of the most harmful GHGs (GHG emissions from landfills are equivalent to 22.9 million passenger vehicles).[110] In economic terms, this loss is greatest at the point where a finished product is discarded and enters a mixed waste stream.

Food waste can have a dramatically varied impact, depending on the amount produced and how it is dealt with. In some countries the amount of food waste is negligible and has little impact; in countries such as the US and the UK, however, the social, economic and environmental impact of food wastage is enormous. The disposal of food waste costs around $1 billion to the US every year, according to the US EPA,[111] and landfills account for 34% of all methane emissions in the US.

A recent estimate (March 2010) says that the US uses about 16% of its energy consumption to produce food (Figure 2.7).[112] This includes its cultivation, transportation, processing, sale, storage and preparation. Overproduction involves food, labour, energy and disposal costs. When food is thrown away, the energy used to produce it is wasted too. The study estimates that the amount of food wasted represents about 2% of annual energy consumption in the US, which is far from trivial.

At the heart of controversies over how food is transported and traded—about 'food miles', 'food mountains' and 'food deserts', for example—are concerns about climate change, as transportation has implications for the whole food system.

Food waste is also water waste, as large quantities of water are used to produce the wasted food. Undoubtedly, agricultural and food production losses are particularly high between field and market in developing countries, and wastage (*i.e.* excess calorie intake and obesity) is highest in the more industrialized nations. The loss of, or reduction in, other primary ecosystem services (*e.g.* soil structure and fertility; biodiversity, particularly pollinator species; and genetic diversity for future agriculture improvements) and the production of GHGs (notably methane) by decomposition

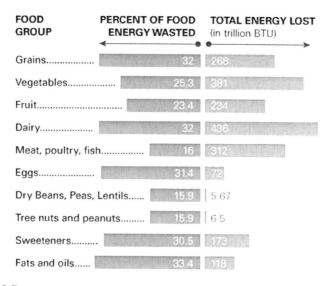

FOOD GROUP	PERCENT OF FOOD ENERGY WASTED	TOTAL ENERGY LOST (in trillion BTU)
Grains..............	32	268
Vegetables...............	25.3	381
Fruit..............................	23.4	234
Dairy..................	32	436
Meat, poultry, fish.............	16	312
Eggs..................	31.4	72
Dry Beans, Peas, Lentils......	15.9	5.67
Tree nuts and peanuts.........	15.9	6.5
Sweeteners.........	30.5	173
Fats and oils......	33.4	118

Figure 2.7   Food and energy waste in America. The left column represents the percentage of food energy wasted in each food group and the right column the total amount of energy lost for each group. Data are from 2004.

of the discarded food, are just as important to long-term agricultural sustainability the world over.

Wasting food is not only an inefficient use of ecosystem services and of the fossil fuel-based resources that go into producing them, but also a significant contributor to global warming once in landfills. An average of at least 1.9 t CO_2 equivalent per tonne of food wasted is estimated to be emitted in Europe during the whole life cycle of food waste.[113] At European level, the overall environmental impact is at least 170 Mt of CO_2 equivalent emitted per year (close to the total GHG emissions of Romania or of the Netherlands in 2008, and approximately 3% of total EU-27 emissions in 2008). This figure includes all steps of the life cycle of food waste: agricultural steps, food processing, transportation, storage, consumption steps and end-of-life impacts.

Considering the performance of respective sectors, the household sector presents the most significant impact, both per tonne of food waste (2.07 t CO_2 equivalent per tonne) and at the European level (78 Mt CO_2 equivalent per year), at 45% of estimated annual GHG emissions caused by food waste. Food waste generated in the manufacturing sector is responsible for approximately 35% of annual GHG emissions.

Limitations of these estimates relate to the reliability of the food waste quantities calculated by each country, as well as to the nature of environmental data available in existing studies. Only environmental data about the food sector in general (production, consumption) in Europe were available and thus used.

The University of Arizona believes that if Americans cut their food waste in half, it would reduce the country's environmental impact by 25%.[114] WRAP, which says the entire food supply chain in the UK contributes 20% of its GHGs, believes that if we stopped throwing out edible food, the impact it would have on CO_2 emissions would be the equivalent of taking 1 in 5 cars off the road.[115]

But ironically, one of the solutions to dealing with food waste actually results in a product that could keep cars on the road: biogas. Biogas is a by-product of a process called anaerobic digestion. This is a process where organic matter, such as food waste, breaks down in an environment with little or no oxygen, generating a natural gas made up of 60% methane and 40% CO_2. It is the exact process, in fact, which goes on in landfills, but there is a difference. Whereas methane can be harmful to the environment in an open setting, such as a landfill, in a controlled and closed setting such as a combined heat and power plant, it can be harnessed and converted into biogas, a renewable energy source that can be used to provide heat, light and fuel. According to a study by the National Society for Clean Air, biogas-fuelled cars can reduce CO_2 emissions by anything from 75% to 200% compared to cars powered by fossil fuels.[116]

Most organic matter can be processed by aerobic digestion. In the UK it is already being used to treat sewage, which according to Friends of the Earth (FOE) reduces CO_2 emissions by 16% compared to traditional sewage treatments.[117] According to the Chartered Institute for Environmental Health, gas from sewage waste and landfills is already being used to provide 650 MW of electricity to the UK's national grid, representing 60–75% of the country's green energy (the UK is Europe's biggest producer of biogas).[118]

However, while the potential for food waste-as-energy seems large, the practical applications for it are currently very small (only 0.4% of the UK's food waste is processed by aerobic digestion, for example), with critics of AD pointing out that the amount food waste can contribute to the energy supply is negligible to say the least.[119] If 5.5 Mt of food waste were treated by anerobic digestion (the majority of the UK's annual 6.7 Mt of food waste) it could generate enough electricity to power 164 000 houses. That said, environmentalists will point out that this is much better than getting that electricity from fossil fuels. And there has been a big push, in Europe in particular, to cut back on the amount of biodegradable waste that is being sent to landfills. According to the European Landfill Directive, the amount of biodegradable waste sent to landfills in Member States by 2020 must reach 35% of the levels reached in 1995.[120]

The country that is leading the way in putting its biodegradable waste mountains to good use, particularly in the area of biogas-powered cars, is Sweden. There are already 7000 biogas cars on the road in Sweden, and the country plans to eliminate petrol and diesel vehicles from the streets by 2020. It also has 779 biogas buses and the world's first biogas train, which, according to *The Ecologist*, cost just €1 million ($1.4 million) to develop.[121]

Changing the perception of waste as something that needs to be disposed of, to one of waste as a commodity with economic and renewable energy value in the agricultural and food production industries, should be encouraged. Governments can provide support and an enabling policy environment in terms of awareness raising, technology innovation and transfer, agricultural extension to farmers, and support policies that foster managing and recycling of agricultural and food production waste into animal feed.

Losses at the farm level are probably about 15–35%, depending on the industry. The retail sector has comparatively high rates of loss of about 26%, while supermarkets, surprisingly, only lose about 1%. Overall, losses amount to around US$90 billion–100 billion a year.[122]

2.7   SOCIO-ECONOMIC IMPACTS OF FOOD WASTE

Apart from damage to the environment, throwing away food has economic and nutritional losses deriving from the calories lost in discarded food as well as from the energy and materials used to transport food waste to landfills. Wasted food means fewer nutrients are available for human consumption, which jeopardizes community food security.

Food waste carries a hefty price tag, estimated at \$150 billion per year on a dollar per pound basis. In today's economic climate, these figures are hard to ignore. This cost is carried all the way from the farm to the fork, where it is ultimately reflected on your tab. Food waste means a considerable waste of money. WRAP current figures suggest each week a typical household throws away between £4.80 and £7.70 of food that could have been eaten; this is equivalent to £250–400 a year or £15 000–24 000 in a lifetime. In the UK treatment of waste can cost as much as £500 per tonne.[123] *Waste Not, Want Not: Feeding the Hungry and Reducing Solid Waste Through Food Recovery*, a joint publication of EPA and USDA, states USA spends about \$1 billion dollars a year to dispose of food waste.[124]

Business schools teach that waste is a sign of inefficiency, market failure and a cost to industry. With the high monetary costs associated with food waste in the supply chain and the social and environmental implications of allowing food to just go to landfill, there could be huge potential savings for the sector.

Currently the amount of surplus food captured and redistributed is unlikely to exceed 15 000 t per year within the UK, according to the UK-based charity FareShare.[125] This amount of recovered and redistributed waste is minimal relative to the scale of the production of food waste. Reasons for a lack of recovery and redistribution of food can partly be put down to worries over food safety and partly to potential brand damage from food finding its way back on to the market.

The Food Waste Management Cost Calculator—a tool which estimates the cost competitiveness of alternatives to food waste disposal, including source reduction, donation, composting, and recycling—is available on the internet.[126] The calculator demonstrates that environmentally and socially responsible food waste management is cost-effective for many facilities and waste streams.

REFERENCES

1. FAO, *Food Loss Prevention in Perishable Crops*, FAO Agricultural Service Bulletin, no. 43, FAO Statistics Division, 1981.
2. T. Stuart, *Waste - Uncovering the Global Food Scandal*, Penguin, 2009.

3. V. Smil, *Improving Efficiency and Reducing Waste in our Food System*, J. Integr. Environ. Sci., 1, 2004, 17–26.
4. http://www.oecd.org/dataoecd/30/7/2400395.pdf
5. http://eur-lex.europa.eu/LexUriServ/LexUriServ.do?uri= CELEX:31975L0442:EN:HTML
6. http://eur-lex.europa.eu/LexUriServ/LexUriServ.do?uri= CELEX:31991L0156:EN:HTML
7. http://www.rockandwrapitup.org/attachments/article/103/ TermPaper.pdf
8. WRAP, *Household Food and Drink in the UK*, November 2009 available at http://www.wrap.org.uk/downloads/Household_ food_and_drink_waste_in_the_UK_-_report.ee6f0c12.8048.pdf
9. W. Kling, *Food Waste in Distribution and Use*, Journal of Farm Economics Vol. 25, No. 4, pp. 848–859, (Nov., 1943).
10. W. Kling, 1943.
11. http://www.fao.org/docrep/meeting/007/F5340E/F5340E03. htm
12. http://www.fao.org/docrep/t0073e/t0073e01.htm
13. M. Grolleaud, *Post-Harvest Losses: Discovering the Full Story. Overview of the Phenomenon of Losses during the Post-Harvest System*. Rome, Italy: FAO, Agro Industries and Post-Harvest Management Service, 2002.
14. J. Lundqvist, C. de Fraiture and D. Molden, *Saving Water: from Field to Fork—curbing Losses and Wastage in the Food Chain*, in SIWI Policy Brief. Stockholm, Sweden, 2008.
15. M. Griffin, J. Sobal and T. Lyson, *An Analysis of a Community Food Waste Stream*, Agric. Hum. Values, 2009, 26, pp. 67–81.
16. http://www.mottainai.info/english
17. K. Masuda, *Kenkyusha's New Japanese-English Dictionary*, Kenkyusha Ltd, 1974, p. 1139.
18. J. Scanlan, *The World Turned Inside Out—Waste in History and Culture*, Cambridge Scholars Publishing, 2009.
19. C. T. Anderson, *Sacred Waste: Ecology, Spirit, and the American Garbage Poem*, Interdisciplinary Studies in Literature and Environment 17, 2010, p.35.
20. D. Lee Brien, *From Waste to Superbrand: The Uneasy Relationship between Vegemite and Its Origins*, M/C Journal,

Vol. 13, No. 4, 2010, available at http://journal.mediaculture. org.au/index.php/mcjournal/article/viewArticle/245

21. C. J. M. Ondersteijn, *Quantifying the Agri-food Supply Chain*, Springer 2006.

22. T. H. Davenport, *Process Innovation: Reengineering Work through Information Technology*, Boston, MA: Harvard Business School Press, 1993.

23. J. G. A. J. Vorst, S. J. Dijk and A. J. M. Beulens, *Leagile Supply Chain Design in Food Industry; an Inflexible Poultry Supply Chain with High Demand Uncertainty, the International Journal on Logistics Management*, Vol. 12, No. 2, pp. 73–85, 2001.

24. Lundqvist *et al.*, 2008.

25. M. Grolleaud *et al.*, 2002.

26. T. S. Jayne, A. Chapoto, I. Minde and C. Donovan, *The 2008/09 Food Price and Food Security Situation in Eastern and Southern Africa: Implications for Immediate and Longer Run Responses*, International Development Working Paper #96, Michigan State University, East Lansing, 2009.

27. T. Reardon, C. B. Barrett, J. A. Berdegue and J. Swinnen, *Transformation of Agrifood Systems and Small Farmers in Developing Countries*, World Development, Special Issue, forthcoming, introductory article.

28. K. Chen, A. W. Shepherd and C. da Silva, *Changes in Food Retailing in Asia: Implications of Supermarket Procurement Practices for Farmers and Traditional Marketing Systems*, FAO Agricultural Management, Marketing and Finance Occasional Paper No. 8. Rome: Food and Agricultural Organization, 2005.

29. K. Chen, *et al.*, 2005.

30. P. W. Bettoli and G. D. Scholten, *By Catch Rates and Initial Mortality of Paddle Fish in a Commercial Gillnet Fishery* , Fish Res, 77, pp. 343–347, 2006.

31. FAO, *The State of World Fisheries and Aquaculture*, 2009 available at http://www.fao.org/docrep/011/i0250e/i0250e00. HTM

32. FAO, *Code of Conduct for Responsible Fisheries*, FAO, Rome, p. 41, 1995.

33. A. Kader, *Increasing Food Availability by Reducing Post-harvest Losses of Fresh Produce*, 2005 available at http://postharvest.ucdavis.edu/datastorefiles/234–528.pdf

34. R. A. Cappellini and M.J. Ceponis, *Postharvest Losses in Fresh Fruits and Vegetables*, in: H. E. Moline, (ed.) Post-harvest Pathology of Fruits and Vegetables: Postharvest Losses in Perishable Crops, Univ. Calif. Bull, pp. 24–30, 1984.

35. S. Henningsson and M. Smith, *Waste Minimization in the Food and Drink Industry*, Environmental Protection Bulletin 069, Institute of Chemical Engineers, Journal, pp. 3–9, 2000.

36. Grida Publications, *World Food Supply Food from Animal Feed*, available at http://www.grida.no/publications/rr/food-crisis/page/3565.aspx

37. European Commission (DG ENV) Directorate C—Industry, *Preparatory Study on Food Waste across EU 27*, available at http://ec.europa.eu/environment/eussd/pdf/bio_foodwaste_report.pdf

38. K.W. Stiegert and D. H. Kim, *Structural Changes in Food Retailing: Six Country Case Studies*, FSRG Publication, November 2009, available at http://www.aae.wisc.edu/fsrg/publications/Monographs/!food_retailingchapter7.pdf

39. T. Lang, *Food Industrialization and Food Power: Implications for Food Governance*, Development Policy Review 21, no. 5–6, pp. 558, 2003.

40. T. Lang, *Food Industrialization and Food Power: Implications for Food Governance*, Development Policy Review 21, no. 5–6, p. 560, 2003.

41. P. Dobson, *Buyer Power in Food Retailing: the European Experience*, Paper presented at OECD Conference on 'Changing Dimensions of the Food Economy', The Hague, 6–7 February 2003.

42. G. Rayner, D. Barling and T. Lang, *Sustainable Food Systems in Europe: Policies, Realities and Futures*, Journal of Hunger & Environmental Nutrition, 3:2, p. 158, 2008.

43. G. Rayner, D. Barling and T. Lang, *Sustainable Food Systems in Europe: Policies, Realities and Futures*, Journal of Hunger & Environmental Nutrition, 3:2, p. 159, 2008.

44. http://www.igd.com

45. *The Food & Grocery Monitor*—Issue Two, Report 2011 available at http://www.igd.com/index.asp?id=1&fid=2&sid=2&tid=92

46. http://www.capgemini.com/services-and-solutions/by-industry/retail/overview/

47. http://www.publications.parliament.uk/pa/cm200910/cmselect/cmenvfru/230/23008.htm

48. B. Alfaro, *Application of 'Fish Shelf Life Prediction (FSLP)' Software for Monitoring Seafood Quality in the Cold Chain*, Paper presented at the International Workshop 'Cold-Chain Management', Bonn, Germany, 2 June 2008.

49. S. Kraiselburd, *When is Vendor Managed Inventory Good for the Retailer? Impact of Relative Margins and Substitution Rates*, Instituto de Empresa, Business School Working Paper No. WP06–18, February 23, 2006, available at http://ssrn.com/abstract=1015571

50. http://www.news-medical.net/.../TimeTemps-shelf-life-indicator-for-food-will-help-Norwegian-retailers-reduce-waste.aspx

51. http://www.timetemperature.com/tzus/time_zone.shtml

52. http://en.wikipedia.org/wiki/Garbology

53. http://www.enotes.com/food-encyclopedia/food-waste

54. http://metamedia.stanford.edu:3455/17/174

55. http://www.todaysdietitian.com/newarchives/tdsept2007pg84.shtml

56. WRAP, *Research into Consumer Behaviour in Relation to Food Dates and Portion Sizes*, 2008, available at http://www.wrap.org.uk/retail_supply_chain/research_tools/research/report_food_waste.html

57. http://ec.europa.eu/environment/eussd/pdf/bio_foodwaste_report.pdf

58. WRAP, *Household Food and Drink Waste in the UK*, United Kingdom, November 2009.

59. T. Jones, *The Value of Food Loss in the American Household*, Bureau of Applied Research in Anthropology, A Report to Tilia Corporation, San Francisco, CA, USA, 2004.

60. WRAP, *Down the Drain: Quantification and Exploration of Food and Drink Disposed of to the Sewer by Households in the UK*, (WRAP Project EVA063), United Kingdom, 2009.

61. S. J. Van Garde and M. J. Woodburn, *Food Discard Practices of Householders*, J. Am. Diet. Assoc., 87, pp. 322–329, 1987.

62. B. Lyndhurst, *Food Behaviour Consumer Research—Findings from the Quantitative Survey*, Briefing Paper. UK: WRAP, 2007.

63. R. Wenlock, D. Buss, B. Derry and E. Dixon, *Household Food Wastage in Britain*, Br. J. Nutr. 43, pp. 53–70, 1980.

64. R. Osner, *Food Wastage*, Nutrition and Food Science, 13–16. July/August 1982.

65. C. Hamilton, R. Dennis and D. Baker, *Wasteful consumption in Australia. Discussion Paper Number 77, March 2005. Manuka, Australia: The Australia Institute*. ISSN 1322–5421

66. T. W. Jones, *Using Contemporary Archaeology and Applied Anthropology to Understand Food Loss in the American Food System*, Bureau of Applied Research in Anthropology, University of Arizona available at http://www.ce.cmu. edu/gdrg/readings/2006/12/19/Jones_UsingContemporary ArchaeologyAndAppliedAnthropologyToUnderstandFood LossInAmericanFoodSystem.pdf

67. J. Lundqvist *et al.*, 2008.

68. T. Jones, *Addressing Food Wastage in the US*. Interview: The Science Show, 8 April 2006 available at http://www.abc.net.au/rn/scienceshow/stories/2006/1608131.htm

69. R. Sibriá, J. Komorowska and J. Mernies, *Estimating Household and Institutional Food Wastage an Losses in the Context of Measuring Food Deprivation and Food Excess in the Total Population*, Statistics Division, Working Paper Series no: ESS/ESSA/001e. Rome, Italy: FAO, 2006.

70. K. D. Hall, J. Guo, M. Dore and C. Chow, *The Progressive Increase of Food Waste in America and its Environmental Impact*, PLoS ONE, 2009.

71. http://www.epa.gov/osw/rcc/resources/meetings/rcc-2010/schwab.pdf

72. http://foodpsychology.cornell.edu/about/about.html

73. BIO IS, *Preparatory Study on Food Waste across EU 27*, December 2009–September 2010.

74. *WRAP* (a) *Household Food and Drink Waste in the UK*, Banbury, UK, 2009. *WRAP* (b) *Down the Drain: Quantification and Exploration of Food and Drink Waste disposed of to the Sewer by Households in the UK*, Banbury, UK, 2009.

75. SEI, Saastva Esti Instituut, *Eestis Tekkinud Olmejäätmete (Sh Eraldi Peakendijäätmete Ja Biolagunevate Jäätmete) Koostise Ja Koguste Analüüs*, Talinn, Estonia, 2008 and in European Commission [DG ENV—Directorate C], Final Report—Preparatory Study on Food Waste, October 2010.

76. ADEME, ACR +, ARC, LIPOR, *Dossier de Presse: Semaine Européenne de la Réduction des Déchets*, Paris, France, 2009 and in European Commission [DG ENV—Directorate C] Final Report—Preparatory Study on Food Waste, October 2010.

77. Irish Environmental Protection Agency, *2008 National Waste Report and European Commission*, 2009 and in European Commission [DG ENV—Directorate C] Final Report—Preparatory Study on Food Waste, October 2010.

78. Danish Environment Ministry, *Food Waste Report* and in European Commission, 2010 and in European Commission [DG ENV—Directorate C] Final Report—Preparatory Study on Food Waste, October 2010.

79. Naturvardsverket, *Personal Communication on Waste Generation*, Stockholm, Sweden, 2010 and in European Commission [DG ENV—Directorate C] Final Report—Preparatory Study on Food Waste, October 2010.

80. WRAP, *Love Food Hate Waste,* 2010 available at www.lovefoodhatewaste.com/about_food_waste

81. E. P. Cathcart and A.M. Murray, *A Note on the Percentage Loss of Calories as Waste on Ordinary Mixed Diets, J. Hyg.* 39, p. 45, 1939.

82. R. Wenlock and D. Buss, *Wastage of Edible Food in the Home: a Preliminary Study*, J. Hum. Nutr. 31, pp. 405–411, 1977.

83. R. Osner, *Food Wastage*, Nutrition and Food Science, 13–16. July/August, 1982.

84. WRAP, *The Food We Waste,* Banbury, UK, 2008 and WRAP, *Household Food and Drink Waste in the UK,* Banbury, UK, 2009.

85. WRAP, *A Reiew of Waste Arising in the Supply of Food and Drink to UK Households,* Banbury, UK, 2010.

86. http://www.wrap.org.uk/downloads/Dealing_with_Food_Waste_-_Final_-_2_March_07.f11ca879.3603.pdf

87. Tech Innovation Ltd. 2004, *United Kingdom Food and Drink Processing Mass Balance: A Biffaward Programme on Sustainable Resource Use*, available at http://www.massbalance.org/downloads/projectfiles/2182–00335.pdf

88. WRAP, 2010.

89. WRAP, 2010.

90. A. Segrè and L. Falasconi, *Il Libro Nero dello Spreco*, Edizioni Ambiente, Italy, 2010.

91. A. Segrè, *Elogio dello -SPR + ECO, Formule per una Società Sufficiente*, Arianna Editrice, Italy, 2008.
92. www.adoc.org/notizie/5312/sprechi-alimentari-2011
93. http://fciweb.nic.in
94. Rediff. Com, *How much Food does India Waste?* 17 March 2007, available at www.rediff. com/cms/print/jsp?docpath = money//2007/mar/16food.htm cited in J. Lundqvist, C. de Fraiture and D. Molden. *Saving Water: From Field to Fork— Curbing Losses and Wastage in the Food Chain*, SIWI Policy Brief, SIWI, 2008.
95. E. Morgan, *Fruit and Vegetable Consumption and Waste in Australia*, State Government of Victoria, Victorian Health Promotion Foundation, 2009.
96. S. Yoon and H. Lim, *Change of Municipal Solid Waste Composition and Landfilled Amount by the Landfill Ban of Food Waste*, J. KORRA 13, pp. 63–70, 2005.
97. W. Baek, *Change of MSW Composition Attributed by Ban on Direct Landfill of Food Waste in Korea*, Presentation to the 7th Workshop on GHG Inventories in Asia (WGIA7), 7 July 2009. South Korea: Environmental Management Corporation.
98. http://www.japanfs.org/en/mailmagazine/newsletter/pages/027774.html
99. http://www.env.go.jp/recycle/low-e.html
100. L. Kantor, *A Dietary Assessment of the U.S. Food Supply: Comparing Per Capita Food Consumption with Food Guide Pyramid Serving Recommendations*, Report no. 772. Washington, DC: Food and Rural Economics Division, Economic Research Service, USDA, 1998.
101. M. Griffin *et al.* 2009.
102. T. Jones, 2004.
103. T. Jones, 2006.
104. USEPA, *Municipal Solid Waste Generation, Recycling, and Disposal in the United States: Facts and Figures for 2008* (November 2009), available at http://www.epa.gov/epawaste/nonhaz/municipal/pubs/msw2008rpt.pdf
105. http://www.cdc.gov/nchs/fastats/overwt.htm
106. J. Lundqvist *et al.* 2008.
107. http://www.fao.org/tc/tca/work05/CIRAD.pdf
108. http://www.fao.org/ag/againfo/themes/en/dairy/pfl/home.html

109. BIO IS, *Preparatory Study on Food Waste across EU 27*, December 2009–October 2010 available at http://ec.europa. eu/environment/eussd/pdf/bio_foodwaste_abstract.pdf
110. http://www.statistics.gov.uk/downloads/theme_environment/ transport_report.pdf
111. Food Waste Management Cost Calculator available at http:// www.epa.gov/osw/conserve/materials/organics/food/tools/ foodcost.pdf
112. http://www.enn.com/agriculture/article/41848
113. European Commission (DG ENV) Directorate C—Industry, *Preparatory Study on Food Waste across EU27*, available at http://ec.europa.eu/environment/eussd/pdf/bio_foodwaste_ report.pdf
114. http://www.eateco.org/Waste.htm
115. WRAP, *A Review of Waste Arisings in the Supply of Food and Drink to UK Households*, Banbury, UK, 2010.
116. http://www.ecowatch.org/pubs/junjul08/whyis.htm
117. http://www.foe.co.uk
118. http://www.cieh.org
119. http://www.foe.co.uk/resource/briefings/food_waste.pdf
120. *All About: Food Waste* by R. Oliver, All About: Food Waste available at http://alyakenglish.com/pds/pds/attach_file_01/ All%20About%20Food.doc
121. http://news.bbc.co.uk/2/hi/science/nature/4373440.stm
122. T. Jones, 2006.
123. http://www.igd.com/index.asp?id = 1&fid = 1&sid = 17&tid = 0&folid = 0&cid = 165
124. http://www.epa.gov/osw/conserve/materials/organics/pubs/ wast_not.pdf
125. FareShare, *Information about Operations and Benefits of FareShare Redistribution of Food*, 2008 available at http:// www.fareshare.org.uk/about/index.html
126. http://www.epa.gov/osw/conserve/materials/organics/food/tools/
127. http://www.wfp.org/content/annual-report-2009
128. http://www.weforum.org/issues/global-competitiveness
129. A. Kader, *Perspective on Postharvest Horticulture, HortScience 38*, pp. 1004–1008, 2003.
130. C. Nellemann, *The Environmental Food Crisis. Kenya: United Nations Environment Programme*, Nairobi, Kenya, 2009.

CHAPTER 3

Food Waste in Food Policies and Legislation: a Controversial Issue

Sustainable food consumption can be defined as access and use by all present and future generations of the food necessary for an active, healthy life through means that are economically, socially and environmentally sustainable.

Lefin (2008)

3.1 GLOBAL FOOD POLICIES

The food system, as currently conceptualized, embraces the full length of the food chain and its interactions in its wider socio-economic and environmental settings where a variety of interests seek to shape the inputs and outputs of the food chain.

Global food systems and consequently food policies and legislation, are in constant evolution as culture, consumption patterns and needs are continuously developing.[1] In this book we use the term 'food policy' to refer to any policy that addresses, shapes or regulates the decision-making environment of food producers, food consumers and food marketing agents in order to further social objectives. Such objectives nearly always include improved and safer nutrition for citizens, growth in domestic food

Transforming Food Waste into a Resource
By Andrea Segrè and Silvia Gaiani
© Andrea Segrè and Silvia Gaiani, 2012
Published by the Royal Society of Chemistry, www.rsc.org

production, equal income-earning opportunities and security against famines and other food shortages.

Food policy covers a wide range of policy fields and involves a variety of actors who have different interests and aims, for example:

- **Producers**—agricultural actors, food industry and their organizations, including trade unions who are mainly interested in costs and sales.
- **Trade, distributors and their organizations**—whose concerns lie within quantities, availability, logistics and sales.
- **Consumers**—whose main concerns are availability, health and costs.
- **Politicians and administrations**—whose concern lie in the ability to provide efficient policy responses to public and private stakeholders' issues.

Food policy-making and regulation are problematic.[2] After the Second World War food policy used to concern mainly famine and food insecurity, but it now encompasses a whole range of complex issues such as obesity; food safety, competition policy in the retail sector, environmental challenges connected to water, energy, soil management, biodiversity loss and land use competition, pressing social issues such as labour and skills shortages, food quality problems, prices and waste. As a consequence, the links between production, marketing and consumption are complex and, as always, there are far too many interests in play.

Few countries have a coherent food and nutrition policy, or even a coherent capacity to make food policy, partly due to the cross-cutting nature of food policy and partly due to issues that cut across national borders. Most countries start from a food price policy environment that uses food imports and budget subsidies for across-the-board consumer protection, while a host of production-oriented government projects attempts to increase food output.

At national level, food policy has historically been the preserve of ministries of agriculture, with a supporting role played by ministries of health and, in some places, departments dealing with drought relief and rehabilitation. Increasingly the actors are shifting, with more involvement from ministries of trade and industry, ministries of the environment, competition authorities, and even ministries of finance. The EU, and many of its Member States, have for instance created independent food standards agencies[3] which regularly issue guidance to food industry representatives and other stakeholders on a range of topics, often as a result of new regulations coming into force.

Internationally, there is evidence that the new food policy is driving changes to the agenda of organizations like the United Nations Food and Agriculture Organization (FAO) and that the role of international regulatory bodies like the Codex Alimentarius[4] has expanded significantly, for example to assess food safety risks associated with the new biotechnology.

Since 1996, when the World Food Summit called by the United Nations took place in Rome and 185 nations signed a commitment to halve the number of hungry people by 2015,[5] the elimination of food poverty has become a central theme of many policy statements by most of the development institutions. The simultaneous persistence of widespread extreme food deprivation and plentiful food supplies in a world with modern means of communications and transportation suggests that there are fundamental flaws in the ways in which nations function and are governed.

The World Health Organization (WHO) has appealed to governments to act to prevent the double burden of food-related ill-health problems associated with under- and overconsumption coinciding in the same country.[6] WHO and FAO advocate for the end of the productionist era in food policy in favour of better trade and distribution policies. Simply to increase the quantity of food on the market is not an adequate policy goal. Quality, distribution, and externalized social, health and environmental costs also have to be central to the food policy framework.

To develop a coherent and sustainable food strategy, the 2002 World Summit on Sustainable Development (WSSD) in Johannesburg urged for 'a change in the unsustainable patterns of production and consumption',[7] calling for a holistic approach to minimize negative environmental impacts along the food supply chain from production to consumption while at the same time assuring a healthy diet. Governments at the Johannesburg summit called for the development of a 10-year framework of programmes (10-YFP)[8] in support of regional and national initiatives to accelerate the shift towards sustainable consumption and production

(SCP) patterns and pointed out that a piecemeal reform could produce an overly complex and unsatisfactory set of agreements, rules, institutions and programmes.

Efforts by national and international political institutions to assure sound food policies are hampered by the complexities of food policies themselves. Lang[9] identifies public pressure as one of the main drivers of policy change in the food arena, reflecting concerns about health and the state of the environment. Public pressure is beginning to mount. Food activism is growing fast, both in developed countries (*e.g.* the UK Food Group[10]) and internationally (*e.g.* the food sovereignty movement[11]).

During the past period of crises (1980–2000) public pressure drew attention on the unnecessary use of food additives, the impact of pesticides, weak microbiological standards (particularly for food-borne pathogens), limited and unclear labelling, and the role of diet in degenerative diseases such as heart disease, diabetes and some cancers. Social disapproval, mass movements and interest-group action, as well as the core medical arguments, have been the triggers for public action on obesity in the United States and other Western countries.

The capacity to respond to public pressure and to make better food policy depends on the ability to find answers and solutions to the new and old 'fundamentals' of the 21st century: climate change, a fuel/oil/energy squeeze, water stress, competition over land use, labour pressures, urbanization, population increase, dietary change and the nutrition transition with accompanying healthcare costs, social inequalities within and between countries,[13] and the burning issue of food waste. (It is also important here to remember that food policy is expensive: for example, the Food Standards Agency in the UK has a staff of 600 people and an annual budget of £115 m.[12])

Table 3.1 is an attempt to represent the 'old' or 'new' characterizations of the food policies: most countries are in between and are moving along a continuum from old to new. The changes have many causes and three main sets of issues can be identified:

- the character of the food system
- the effects on the human population
- the actors and agendas of food policy.

Table 3.1   Food policy old and new

	Food policy 'old'	*Food policy 'new'*
1. Population	Mostly rural	Mostly urban
2. Rural jobs	Mostly agricultural	Mostly non-agricultural
3. Employment in the food sector	Mostly in food production and primary marketing	Mostly in food manufacturing and retail
4. Actors in food marketing	Grain traders	Food companies
5. Supply chains	Short - small number of food miles	Long - large number of food miles
6. Typical food preparation	Mostly food cooked at home	High proportion of pre-prepared meals, food eaten out
7. Typical food	Basic staples, unbranded	Processed food, branded products More animal products in the diet
8. Packaging	Low	High
9. Purchased food bought in	Local stalls or shops, open markets	Supermarkets
10. Food safety issues	Pesticide poisoning of field workers Toxins associated with poor storage	Pesticide residue; in food; Adulteration Bio-safety issues in processed foods (salmonella, listeriosis)
11. Nutrition problems	Under-nutrition	Chronic dietary diseases (obesity, heart disease, diabetes)
12. Nutrient issues	Calories, rnicronutrients	Fat; sugar; salt
13. Food-insecure	'Peasants'	Urban and rural poor
14. Main sources of national food shocks	Poor rainfall and other production shocks	International price and other trade problems
15. Main sources of household food shocks	Poor rainfall and other production shocks	Income shocks causing food poverty
16. Remedies for household food shortage	Safety nets, food-based relief	Social protection, income transfers
17. Fora for food policy	Ministries of agriculture, relief/rehabilitation, health	Ministries of trade and industry, consumer affairs, finance; Food activist groups, NGOs

Table 3.1   (*Continued*)

	Food policy 'old'	Food policy 'new'
18. Focus of food policy	Agricultural technology, parastatal reform, supplementary feeding, food for work	Competition and rent-seeking in the value chain, industrial structure in the retail sector, futures markets, waste management, advertising, health education, food safety
19. Key international institutions	FAO, WFP, UNICEF, WHO, CGIAR	FAO, UNIDO, ILO, WHO, WTO

Source: http://www.odi.org.uk/resources/download/1238.pdf

In Table 3.1 we point particularly to the industrialization and globalization of the food system. The food system can no longer be understood simply as a way of moving basic commodities from farm to plate. Today, food is increasingly produced by commercial growers, feeding long and sophisticated supply chains which market often processed and branded products to mainly urban consumers. Many people work in the food industry, but few of them are farmers or farm workers: in developed countries, as few as one in ten.[14] The table documents, in a very simple way, the transformation the food system and food policies have been going through in recent years in developed countries. Lang,[15] in particular, focuses on the importance of 13 changes, ranging from how food is grown and animals reared, to the mass marketing of food brands and the concentration of power in food manufacturing and marketing—a topic which will be extensively analysed in the following chapters.

Table 3.1 shows that food policy is in constant evolution and that the only possible approach to food policy is through a multisectoral and coherent strategy able to bridge the problems at the micro level with those at the macro level, the duality between the state systems of regulations and a system of self-regulation, largely driven by the major forces in supply chain management, the food retailers, and in particular the gap between short-run and long-run effects of macro policy.[16]

The above-mentioned food policy aspects relate mainly to developed, Western countries. Developing countries have to deal with a different set of food policy issues, such as food security—including issues such as hunger and malnutrition—food subsidies, food aid and the sufficiency of food supplies during emergencies. Latin America, Asia and Africa suffer from significant deficiencies in the availability and quality of food security data, and gaps remain in our understanding of the complex interrelations between food policy, nutrition and the response mechanisms of individuals, households and markets to changing economic environments. Furthermore, the scarcity of expertise and of finance and unfair trade regulations are significant constraints.

Developing countries are still not effectively participating in the international standard-setting organizations recognized by the World Trade Organization (WTO); they are still weak in negotiating market access and in ensuring that internationally agreed-upon standards reflect their production conditions. Little has been achieved since the 2005 Hong Kong Ministerial Conference when parties agreed to eliminate agricultural export subsidies by 2013 and grant the Least Developed Countries free access to OECD markets for at least 97% of agricultural and manufacturing tariff lines.[17] Although global exports increased from 32% to 37% (2000–2007), Africa's share increased only from 2.3% to 2.8%.[18]

Unfavourable trade regulations combined with the marketing of seeds have forced small farmers into financial and food poverty. Many seed providers have developed transgenic seeds that require farmers to buy new seeds every year instead of being able to replant using the seeds produced by their produce. In addition to that, the traditional retail sector, made up of small general stores selling processed foods and dairy products coming from local producers, is declining fast. New supermarket chains,[19] favoured by the liberalization of retail foreign direct investment, are hindering the development of small economies of scale. This has sparked a rural–urban migration to cities, leading to the development of slums and escalating poverty and hunger.

The global food crisis of 2008, which manifested itself by a combination of high and dramatically increasing prices of food (Figure 3.1)—especially staples—coupled with shortages and diminished food stocks, deeply affected the food policy of many

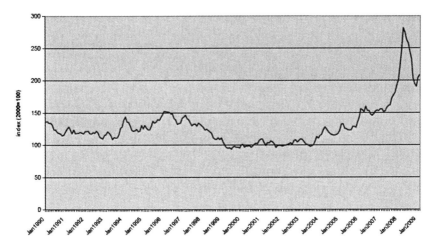

Figure 3.1 Prices, January 1990–February 2009. Source: www.unctad.org/en/
docs/tdbex47d3_en.pdf

developing countries and Africa in particular. Although the situation has somewhat improved, the food security situation of African countries remains worrying. Of the 36 countries worldwide currently facing a food security crisis, 21 are African, and it is estimated that there are now over 300 million Africans facing chronic hunger—nearly one-third of the continent's population.

The 2008 food crisis highlighted the extreme vulnerability of many developing countries' food security to external shocks. High prices and a tightening of supply of basic foodstuffs hit African countries especially hard. This was largely due to the deeper and more protracted production crisis affecting African agriculture. Productivity in African agriculture is low in comparison to other regions of the world. Even more worrying is the fact that this productivity has not seen any real improvement over the past decades.

Another issue which is impacting the food policy in developing countries is the 'land grabbing' phenomenon. One of the most evident effects of the food price crisis of 2007–08 on the world food system has been the proliferating acquisition of farmland in developing countries by other countries seeking to ensure their food supplies (Table 3.2).[20] Food-importing countries with land and water constraints but rich in capital, such as the Gulf States, are at the forefront of new investments in farmland abroad.

Table 3.2   Examples of overseas land investment to secure food policies, 2006–2009

Country investor	Country target	Plot size (hectares)	Current status	Source
Bahrain	Philippines	10 000	Deal signed	Bahrain News Agency, February 2009
China (with private entities)	Philippines	1 240 000	Deal blacked	Fair Trade Alliance, March 2008
China	Zimbabwe	101 171	Deal signed	Chinadialogue, June 2008
Libya	Ukraine	250 000	Deal signed	The Guardian, November 2008
Qatar	Kenya	40 000	Deal signed	Daily Nation, January 2009
United Arab Emirates (with private entities)	Pakistan	324 000	Deal signed	Financial Times, May 2008
South Korea (with private entities)	Sudan	690 000	Deal signed	Korea Times, 2008/09
Saudi Arabia	Tanzania	500 000	Requested	Reuters Africa, April 2009
Jordan	Sudan	25 000	Deal signed	Jordan Times, November 2008

Source: J. von Braun and R. Meinzen-Dick, IFPRI, 2009.

In addition, countries with large populations and food security concerns such as China, South Korea and India are seeking opportunities to produce food overseas. These investments are targeted toward developing countries where production costs are much lower and where land and water are more abundant. Other factors that influence investments include geographic proximity and climatic conditions for preferred staple crops.

In Sudan alone, South Korea has acquired 700 000 ha to produce wheat, and a Chinese company has secured 2.8 million ha for biofuel production. The International Food Policy Research Institute (IFPRI) estimates that 15–20 million ha of land in developing countries is currently under negotiation, although exact figures for how much has already been leased or bought are not known.[21] These land acquisitions have the potential to inject much-needed investment into rural areas in poor developing countries, but they also raise concerns about the impacts on poor local people, who risk losing access to and control over land on which they depend. There is also the risk of creating a form of neo-colonialism.

Developing countries need investments and not a new form of colonialism: they need to be able to participate in the international standard-setting organizations recognized by the WTO, improve their market access and ensure that internationally agreed-upon standards reflect their production conditions.

There is a need for a new global governance architecture of food, agriculture and nutrition: the twinned crises in food and energy are creating a political environment in which real reform of the food system may actually be possible for the first time in a generation.

3.2 GLOBAL FOOD LEGISLATION

Food policies are complex and debated issues, and same can be said of food legislation. Ideally food policies are developed in advance of food legislation, but in some cases it may be the other way around, *i.e.* outstanding policy questions can be clearly identified and resolved in the course of analysing and discussing concrete draft legislative proposals.

Broadly speaking, food laws can be considered policy tools, alongside other instruments such as economic measures, guidelines and other nonbinding instruments and awareness raising and public participation. In the food area, codes on good agricultural practices, good manufacturing practices and good hygiene practices, as well as codes of ethics, have been developed and agreed at international level, and there is burgeoning interest in their implementation.

The term 'food legislation' is generally used to apply to legislation which regulates the production, trade and handling of food. The narrow view would restrict this meaning to the regulation of food control, food safety and food trade at national level, and would focus on laws and regulations that refer to food in general or to specific kinds of food. The broader view would look at the wide variety of fields that must actually be regulated in order to ensure the production, trade and handling of safe food, and would take all of these into account. In other words, everything having to do with food, whether directly or indirectly, would come within the ambit of food law.[22]

Significant regulatory activity has taken place in the international arena with regard to food over the last several years. The Uruguay Round of Multilateral Trade Negotiations in 1994 led to the establishment of the WTO in January 1995.[23] Agriculture was included in the trade talks in a significant way for the first time and it was agreed to reduce tariff barriers for many agricultural products in order to encourage free trade. Two agreements relevant to food, the Agreement on Sanitary and Phytosanitary Measures (SPS Agreement) and the Agreement on Technical Barriers to Trade (TBT Agreement)[24] were concluded within the framework of the WTO. These agreements set important parameters governing the adoption and implementation of food quality and food safety measures.

The TBT Agreement, which had been in existence as a voluntary agreement since the Tokyo Round[25] (1973–1979), was converted into a binding multilateral agreement through the Uruguay Round. The TBT Agreement covers all technical requirements and standards applied to all commodities, such as labelling, that are not covered under the SPS Agreement. The SPS Agreement was drawn up to ensure that countries apply measures to protect human and animal health (sanitary measures) and plant health (phytosanitary measures) based on an assessment of risk, or in other words, based on science. The use of international standards is intended to allow countries to prioritize the use of their often limited resources and to concentrate on risk analysis.

The adoption of the Codex Alimentarius[26] as the source of international food standards by the SPS Agreement in 1995 has been one of the most significant recent influences on food regulation worldwide, and can be seen as an acknowledgment of the increasing globalization of food production and food trade. The objectives of the Codex are to protect the health of consumers, to ensure fair practices in food trade and to promote the coordination of all food standards work undertaken by national governments. Worldwide outbreaks of food-borne disease, with concomitant media attention and outspoken consumer concerns, have in fact triggered unprecedented interest in food regulation and in the infrastructures which govern food safety at international and national level.

For some subject matters, international organizations or arrangements explicitly rely on regional groupings to develop regional standards as well as to discuss international standards and to solicit inputs into their development. For the Caribbean, the Caribbean Community (CARICOM)[27] and the Caribbean Forum of African, Caribbean and Pacific States (CARIFORUM)[28] are the two most significant economic groupings. CARICOM has among its objectives the promotion of economic integration among its Member States, and the establishment of a single market and economy. Similarly, CARIFORUM works toward better coordination of EU support and improved regional integration and cooperation.

For some African countries,[29] the Southern African Development Community (SADC)[30] works toward the development, promotion and harmonization of its Member States' SPS policies. SADC has also established a Food Security Programme, which comprises a number of national and regional projects designed to enhance food security in the region. The programme encourages Member States to implement measures in the medium and long term to increase agricultural productivity and food production, and to promote trade in agricultural commodities.

In South America the Southern Common Market (Mercado Común del Sur, or Mercosur)[31] has led to the establishment of a customs union in 1995 and subsequently a transition phase with a view to constituting the common market. The Member States established an institutional framework, which, in contrast to other regional groupings such as the EU, rejects any notion of supranationality. Food products are supposed to meet the International Codex Alimentarius and the countries have the responsibility to harmonize standards within Mercosur.

Laws, regulations and standards developed at international level cannot be 'imported', as their effectiveness depends on their suitability in specific national contexts. Each country requires policies and legislation tailored to its needs, based on an in-depth analysis of the circumstances in the country, including its existing legislative and regulatory framework for food; policy objectives; institutional capacities; and social, ecological, political and economic conditions. Typically, the legal framework governing food in a particular country reflects a mix of political, societal, economic and scientific forces. Laws and regulations may not have been updated or may have been constantly amended, creating a maze of rules which regulators, industry and consumers find difficult to understand.

The difficulty in many countries is to identify the institutions which are charged with the authority to implement the basic food legislation once it has been amended or enacted. Historically, food control has been considered to be within the purview of the ministry responsible for health (as food safety implicates human health), although certain sectors, such as inspection of meat or other animal products, have traditionally been assigned to the veterinary services.

Some countries may have no food legislation whatsoever, relying solely on international instruments such as Codex standards.[32] Other countries may have comprehensive food legislation but it may be outdated, having been in place for decades. Still others may have religious codes operating in tandem with statutory rules, or may have written policies that are only partially reflected in enforceable and enacted legislation.

3.3 US FOOD POLICIES

Food policy is not an issue American presidents have had to give much thought to, at least since the Nixon administration (1969–1974)—the last time high food prices presented a serious political peril.[33] Since then, federal policies have promoted maximum production of the commodity crops (corn, soybeans, wheat and rice) from which most of the US supermarket foods are derived and have succeeded impressively in keeping prices low and food more or less off the national political agenda.

It must be recognized that the current US food system, characterized by monocultures of corn and soy in the field and cheap calories of fat, sugar and feedlot meat on the table, is not simply the product of the free market. Rather, it is the product of a specific set of government policies that sponsored a shift from solar and human energy on the farm to fossil-fuel energy.[34]

Until the 1920s, agricultural policy in the US was dominated by developmental policy—policy directed at developing and supporting family farms and the inputs of the total agricultural sector, such as land, research, and human labour. After the Second World War, the government encouraged the conversion of the munitions industry to fertilizer—ammonium nitrate being the main ingredient of both bombs and chemical fertilizer—and the conversion of nerve-gas research to pesticides.[35] The government also began subsidizing commodity crops, paying farmers by the bushel for all the corn, soybeans, wheat and rice they could produce.

In the early 1970s, under Secretary of Agriculture Earl Butz, farmers were encouraged to 'get big or get out'.[36] Over the course of the 20th century, farms consolidated into larger, more capital-intensive operations and subsidy policy encouraged these large farms at the expense of small and medium-sized family farms.[37]

Today, approximately 150 000 farmers produce most of the US food and fibre and are among the world's most competitive, able to fully meet domestic needs and also supply large quantities to foreign markets. Although farming itself employs only about 1% of the workforce and accounts for less than 1% of the US gross domestic product (GDP), it is the critical component of the entire food and fibre system—spanning farm inputs, processing, manufacturing, exporting, and a wide range of ancillary services—that contributes $1.5 trillion (16% of GDP) and employs 17% of the labour force.[38]

America's food and agriculture policies include not only ensuring food safety, promoting nutritious and convenient foods and products, delivering food assistance to low-income consumers, protecting environmental quality, and keeping markets functioning, but also labelling and regulation, organic standards, subsidized school lunches, land filling and food waste, food stamps and other hunger-prevention programs such as the federal Woman, Infants and Children (WIC) Nutrition Program,[39] federal subsidies for commodity crops and mandates that international food aid be sourced in the US rather than purchased from local farmers in the global south. These diverse policies fall under the jurisdiction of a number of federal agencies including the Departments of Agriculture, Health and Human Services, Education and the State Department.

The largest piece of federal food policy is the Farm Bill.[40] Renewed every 5 years, the Farm Bill funds a wide range of government programmes including food stamps and nutrition, agricultural research, food safety, animal welfare, forestry, rural electricity and water supply, foreign food aid, and subsidy payments to commodity crop producers. Commodities subsidies cost an average $11.3 billion a year and go substantially to corn, cotton, wheat, rice and soybeans. The main beneficiaries however, are large- and medium-scale commodities farmers and the agribusinesses that supply them. Only 9% of California's 74 000 farms received subsidy payments between 1996 and 2002, while $1.8 billion was paid out to fewer than 3500 farms.[16]

In addition to subsidies, the Farm Bill addresses conservation, nutrition programs, agricultural research and commodity programs. In 2007 the bill amounted to $286 billion in public spending.[41] In the 2008 Farm Bill, 66% of funds were dedicated to nutrition.[42] However, there are loopholes in allocation of funds. Appropriations subcommittees are allowed to approve changes in funding every year, often resulting in cuts to public programs.[43] The lack of funds for these programs limits the effects of the Farm Bill on food security for the average American.

Other federal legislation also affects food policy, including the Schools/Childhood Nutrition Commodity Program (which authorizes funding for school lunches) and the Emergency Food Assistance Program which distributes surplus commodities in addition to purchased goods to food banks and the people they serve.[44]

While the Farm Bill frames the nation's food policy, hundreds of other pieces of federal, state and local regulations also shape how Americans eat. State and local government bodies control much of food policy by deciding where to source food for schools, prisons, hospitals, and government institutions, how to zone cities to create or restrict opportunities for community and urban gardening; and whether or not to invest in developing distribution systems, local food processing facilities and markets for local production and consumption, such as farmer's markets and local food retail.

In recent decades food security, food justice and food sovereignty movements have begun to envision broad, comprehensive approaches to food policy that include all aspects of the food system that involve feeding a population. In this view, food policy is multidisciplinary, multisectoral and intergenerational, and addresses social, political, economic and environmental factors.

In his early study of Food Policy Councils—state coalitions that bring together stakeholders from diverse food-related sectors to examine how the food system is operating and to develop recommendations on how to improve it—Kenneth Dahlberg[45] wrote that a comprehensive food policy should include:

> ... from production issues (farmland preservation, farmers markets, household & community gardens), to processing issues (local *vs.* external), to distribution issues (transportation,

warehousing) to access issues (inner-city grocery stores, co-ops, school breakfasts & lunches, food stamps, the WIC program, *etc.*), to use issues (food safety and handling, restaurants, street vendors), to food recycling (gleaning, food banks, food pantries and soup kitchens) to waste stream issues (composting, garbage fed to pigs, *etc.*).[46]

In addition to this wide variety of topics, a comprehensive food policy should address the interests of the actors involved in the food value chain (farm workers and farmers, processors and workers, retailers and retail workers, *etc.*), as well as those of the consumers. The US fragmented policy approach is not food policy *per se*, but a scattered attempt to address problems in the food system at federal, state and local levels.

Today, a new challenge is represented by the ongoing transformation of US agriculture into the still emerging, global, consumer-driven food system. In the Unites States over 12 000 new food products are introduced annually across 14 major food categories (ranging from baby food to soup).[47] Retail food stores offer choices that provide novelty, variety, and convenience—from organic produce, exotic fruits, and marinated meat to bottled water.

On the other hand a vibrant food justice movement has begun to demand a change in the policies of the current national food system. Driven by this movement, communities and local governments are beginning to look at food policy as essential to meeting the needs of all members of the community. Michael Pollan—professor of journalism at the University of California, Berkeley, author, journalist, activist—advocates that today in America there is soaring demand for local and regional food; farmers' markets, of which the US Department of Agriculture (USDA) estimates there are now 4700,[48] have become one of the fastest-growing segments of the food market. Community-supported agriculture is booming as well: there are now nearly 1500 community-supported farms, to which consumers pay an annual fee in exchange for a weekly box of produce through the season. The local food movement is continuing to grow with no help from the government, and could help the American people to reconnect with the land and consequently put the interests of America's farmers, families and communities ahead of the fast-food industry, the national subsidies and the oil policies.

3.4   US FOOD LEGISLATION

The era of the modern food regulation in the US started in 1906 when President Theodor Roosevelt signed the Pure Food and Drug Act and the Meat Inspection Law.[49] The Pure Food and Drug Act added regulatory functions to the US Bureau of Chemistry. The Meat Inspection Act required the USDA to prevent adulterated livestock from being processed into food.

Food laws continued to evolve based on the concerns of the time. In the 1950s concerns over food additives and pesticides were high. In 1958 the Food Additives Amendment was enacted requiring the evaluation of food additives to establish safety.[50] Throughout the 1980s, various states implemented non-uniform laws to regulate health and nutrition claims. In 1990 the Congress enacted the Nutritional Labeling and Education Act which required nearly all packaged food to bear nutritional labeling.[51]

Currently, regulation of food issues at the federal level is the province of two agencies: the Food and Drug Administration (FDA),[52] whose purposes have been briefly highlighted in the previous paragraphs, and the USDA.[53] The FDA's Center for Food Safety and Applied Nutrition[54] (CFSAN) has jurisdiction over all food, except for meat, poultry, and egg products, which fall under the jurisdiction of USDA's Food Safety and Inspection Service[55] (FSIS). A third agency, the Federal Trade Commission (FTC),[56] has jurisdiction over food advertising. In addition to agencies that regulate at the federal level, each state has its own public health authorities that oversee food safety, and to a lesser extent, food labelling and advertising.

A number of other federal agencies are involved in the direct legislation of food (Table 3.3). Bedrock requirements are set out in statutes enacted by the US Congress and published in the US Code of Federal Regulations. The FDA's statutory authority derives from the Federal Food, Drug, and Cosmetic Act.[57] The USDA's statutory authority derives from the Federal Meat Inspection Act, the Poultry Products Inspection Act, and the Egg Products Inspection Act.[58] The FTC's statutory authority derives from the FTC Act. Pursuant to the authority granted by these statutes, the FDA, USDA, and FTC issue regulations with more detailed requirements. A regulation that is properly issued has the force and effect of law, but federal agencies have increasingly turned to the

Table 3.3   Comparison of US agencies' responsibility for food

Agency	Responsibility
Environmental Protection Agency (EPA)	Drinking-water
	Pesticide residues
Food and Drug Administration (FDA)	Food (but not meat)
	Drug
	Dietary supplements
	Bottled water
	Seafood
	Exotic meat
	Eggs in shell
Federal Trade Commission (FTC) Alcohol and Tobacco Tax and Trade Bureau	Advertising
US Department of Agriculture (USDA)	Alcohol
	Raw vegetables grading
	Raw fruit grading
	Meats
	Poultry
	Eggs, processing and grading

Source: www.usda.gov

issuance of guidance documents to communicate regulatory expectations to industry.

An understanding of how an agency interprets and applies the law also can be gained by studying administrative actions, such as the issuance of warning letters to individual companies, and judicial decisions in cases to which the agency is a party. In addition, advertising disputes are sometimes resolved through the National Advertising Division[59] (NAD) of the Council of Better Business Bureaus, a non-governmental entity. A number of scientists are now calling for an end to this patchwork system and advocate for the creation of a unified food agency.[60]

In term of policy tools, as mentioned before, the Farm Bill is the primary agricultural and food policy tool of the federal government (66% of its budget is dedicated to nutrition). The comprehensive omnibus bill is passed every 5 years by Congress and deals with both agriculture and all other affairs under the purview of the USDA.

State strategies are often discussed and questioned through the Food Policy Councils[61] (FPCs) which consist of representatives and stakeholders from many sectors of the food system. Ideally, the councils include participants representing all five sectors of the food system (production, consumption, processing, distribution and waste recycling). They often include anti-hunger and food justice advocates, educators, non-profit organizations, concerned citizens, government officials, farmers, grocers, chefs, workers, food processors and food distributors. Because they are often initiated by government actors, through executive orders, public acts or joint resolutions, FPCs tend to enjoy a formal relationship with local, city or state officials. The central aim of most FPCs is to identify and propose innovative solutions to improve local or state food systems, spurring local economic development and making food systems more environmentally sustainable and socially just.

3.4.1 US Food Waste Policy and Legislation

The Clinton administration (1993–2001) was the first American administration to get interested in the food waste issue. The secretary of agriculture at the time, Dan Glickman, created a programme to encourage food recovery and gleaning,[62] which means collecting leftover crops from farm fields. He assigned a member of his staff to oversee the programme, and several years were spent in encouraging farmers, schools, hospitals and companies to donate extra crops and food to feeding charities.

On 1 October 1996, President Clinton signed the Bill Emerson Good Samaritan Food Donation Act[63] to encourage the donation of food and grocery products to non-profit organizations for distribution to needy individuals. This Donation Act, still in law, protects donors from civil and criminal liability in case the product donated in good faith later causes harm to the needy recipient. Congress recognized that the provision of food close to recommended date of sale is, in and of itself, not grounds for finding gross negligence. For example, cereal can be donated if it is marked close to code date for retail sale.

Despite all this, in 2005 still less than 3% of food waste was recovered, and as a consequence, more than 97% went to landfills.[64] The US Environmental Protection Agency (EPA) started in the same year to encourage and support the composting of organics and to promote a consistent reduction in food waste.

The framework for the management of waste is now represented by the Resource Conservation and Recovery Act[65] (RCRA). RCRA provides, in broad terms, the general guidelines for the waste management program envisioned by Congress. It includes a congressional mandate directing the EPA to develop a comprehensive set of regulations to implement the law. These regulations, issued by the EPA, translate the general mandate of the law into a set of requirements for the Agency and the regulated community. The RCRA gives EPA the authority to control waste from cradle to grave. This includes the generation, transportation, treatment, storage and disposal of hazardous waste.

EPA is currently implementing many projects, such as the Ohio food waste composting[66] (involving the creation of a tool kit for generators), the Michigan Compost Operators Training[67] (involving the conduction of trainings and developing a tool kit), the Southern Illinois University Vermicomposting project[68] (taking cafeteria waste and feeding it to worms), Eureka Recycling in Minnesota[69] (supporting a proposed residential collection program), and the Ho-Chunk Nation's casino composting[70] (composting casino wastes for use on community gardens).

A food waste composting survey carried out in January 2011[71] found out that throughout the US there are 273 facilities that currently accept food waste. Food waste management strategies are on the whole implemented at the state and local level, so there is not a uniform set of food waste policies across the country (Table 3.4).

Seattle and San Francisco are leading the way in terms of food waste recycling. Since 2009 their new city law requires residents to compost food waste. But while Seattle exempts businesses, restaurants and apartment buildings from the law, San Francisco mandates that all residents, plus businesses, restaurants and multiple occupation units like apartment houses, compost waste. San Francisco then sells the compost to Bay Area farms and vineyards. This programme is the latest effort in one of the most aggressive recycling campaigns in the nation. San Francisco is currently keeping 72% of its garbage stream out of landfill.[72] The city's ultimate and fairly lofty goal is to get to zero waste by the year 2020.

The term 'zero waste' includes 'zero solid waste', 'zero hazardous waste', 'zero toxics' and 'zero emissions'. Zero waste suggests that the entire concept of waste should be eliminated and waste should be thought of as a 'residual product' or simply a 'potential resource'. Opportunities such as reduced costs, increased profits,

Table 3.4 Summary of food waste composting facilities

Region	Total	Greater than 5000 Mg/y	Greater than 50,000 Mg/y	Commercial or municipal composters	Accept residential waste
New England[a]	51	9	2	16	8
Northeast/ Mid- Atlantic[b]	48	6	3	15	3
Southeast[c]	18	4	2	11	3
Upper Midwest[c]	48	13	3	17	10
Mountain[d]	36	6	5	27	13
West[e]	72	19	9	45	34
Entire US	273	57	24	131	71

Source: C. Olivares, N. Goldstein, Food composting infrastructure, BioCycle 49(8), 2008 pp. 34–35.

and reduced environmental impacts are found when returning these 'residual products' or 'resources' as food to either natural and industrial systems.

The zero waste strategy[73]—promoted by Paul Connett, a professor of chemistry at St Lawrence University—considers the entire life-cycle of products, processes and systems in the context of a comprehensive systems understanding of human interactions with nature and search for inefficiencies at all stages. With this understanding, wastes can be prevented through designs based on full life-cycle thinking.

Zero waste may seem more of a platonic ideal than a realistic objective,[74] but a growing number of communities and businesses worldwide are adopting its principles, drawn to both its environmental and economic advantages. Corporations such as Wal-Mart, Nike, and Ford have all set zero waste targets for their operations, and so have the cities of Oakland, San Francisco, and Seattle, among others. Outside the US, New Zealand and regions of Australia and Canada have committed themselves to the zero waste challenge.

3.5 EUROPEAN FOOD POLICIES

In spite of a certain degree of homologation between American and European food production and policy models, there are important differences.[75] In terms of agricultural production the US can be considered a natural exporter.[76] American agriculture is extensive, large in scale, has low production costs due to favourable climates and soils, and serves a sparse population. In contrast, Europe is characterized by a very dense population which has suffered from food shortages in history. The European agricultural sector has evolved in close interrelation with society and, thanks to a corporatist model of intermediation, the EU Member States have safeguarded the interests both of small and large farmers (there are more than 10 million farmers in the EU, but only 2.7 million in the US).[77]

Farming in the US is based on large-scale production and on commodities, whereas the EU model is labour-intensive and made up mainly of small-scale structures. The EU Common Agricultural Policy (CAP) was developed in order to assure food security and enhance agricultural productivity after the Second World War. Since its establishment in 1957 it has undergone more changes than any other EU policy. Since the days of Sicco Mansholt,[78] a former European Commissioner who concentrated all his efforts on modernization, economies of scale and the subsidizing of agricultural products, European food policy has tried to adjust to new demands: oversupply, labour policy, prices, fiscal burden, expansion of the EU and, above all, globalization thorough the GATT agreements whose aim was to encourage free trade between Member States by regulating and reducing tariffs on traded goods and by providing a common mechanism for resolving trade disputes.

In Europe, more than in the US, food policies are called upon to deal with socio-economic complexities, including the cultural dimensions of consumption, the short- and long-term public health considerations of consumption, and different nationally and regionally situated consumers. Changes along the whole food supply chain, as well technical and socio-economic transformations, which have taken place in the last 20 years have forced governments and the EU to respond differently, taming rather than forcing the pace and scale of change.

The EU food system is in a loop,[79] and as a consequence, Member States and the EU institutions are caught in a policy dilemma: on the one hand, they have to actively promote the development of efficient modern food supply chains; on the other hand, they have to develop processes of food governance which can respond to and retain public trust in food. A more environmentally friendly and internationally more acceptable European food system is tenable. Amin argues that a 'shift in emphasis from the politics of place to a politics in place' is urgently required.[80]

Some scientists[81] argue that the CAP is no longer as effective as it used to be. It should be made simpler and more sustainable, with 50% funding for looking after the land and 50% for promoting more sustainable measures. Instead of paying for 'good agricultural practice', the CAP could be used to drive what people need and assure sustainable farming practices in the future. Subsidies could be allocated to promote more sustainable food and farming—healthier for people and the planet. The original CAP of the 1950s was driven by food security policy: now there is a need for a food sustainability policy which could support the development of fair, inclusive, transparent and sustainable food systems.

All people should have access to healthy, safe and nutritious food. The ways in which food is grown, distributed, prepared and eaten should celebrate Europe's cultural diversity,[82] because if food is not produced, processed, and distributed equitably, and if food cultures are irrevocably damaged by product marketing, food can become a vehicle for social discord, inequality and worsening patterns of health.

Europe's dilemma is some ways similar to the American dilemma: how to balance food production for large populations accustomed to unparalleled choice and cheapness with sustainability in both natural and human-ecological terms, thus managing supply chains (including their psychosocial dimensions), surpluses and waste in a manner that enables both them and the earth to sustain future generations.

On 18 November 2010 the European Commission published a Communication (COM(2010) 672 final) on 'the Common Agricultural Policy (CAP) towards 2020—Meeting the food, natural resources and territorial challenges of the future'.[83] The reform aims at making the European agriculture sector more dynamic, competitive and effective in responding to the Europe 2020 vision of stimulating sustainable growth, smart growth and inclusive growth. The EU Agriculture and Rural Development Commissioner, Dacian Cioloş, underlined the importance of viable food production (the provision of safe and sufficient food supplies, in the context of growing global demand, economic crisis and much greater market volatility to contribute to food security) alongside the sustainable management of natural resources and rural areas, and climate action.

The need for SCP had been previously underlined in the EU Sustainable Development Strategy (2006)[84] whose priorities are to facilitate agreement on uniform and scientifically reliable environmental assessment methodologies for food products, the development of an EU Eco Label—which might include food from 2011—the improvement of supply chains' environmental efficiencies and the raising of consumer awareness, raising public procurement for a better environment (COM(2008) 400/2).

The CAP should deliver high standards of environmentally friendly land management, animal welfare and public health in a sustainable way. Sustainability is the new 'umbrella' concept; it shares vision and delivery at EU, national and local levels and integrates currently fragmented policy issues. Future food policies should be based on new indicators (land footprint, greenhouse gases, nutrition costs, localness, seasonality and biodiversity) but the main issues remain how to set policies priorities and whether the EU policy institutions are fit for purpose.

The EC has mainly 'soft' powers on health and consumption (labels, information, education) but stronger powers on the environment. CAP—despite its huge budget of €57 780 438 369 in 2009[85]—should be modernized for ecological public health and the new focus should be on the supply chain as well as farming, competitiveness and agricultural diversity.

3.6   EUROPEAN FOOD LEGISLATION

European Community food legislation was largely conceived as a set of rules prompted mainly by the desire to eliminate trade obstacles within the European internal market. Its genesis can be traced from 1962 to the mid-1980s. During this foundational period, the EC, animated by the goal of establishing an internal market for foodstuffs, pursued a detailed harmonization program consisting of the adoption of directives setting up compositional standards for individual foods. In accordance with the 'traditional approach' to harmonization, the EC prepared nearly 50 'vertical' directives aimed at establishing compositional standards for individual foods, the so-called 'recipe laws'.[86]

This total harmonization approach to food law was not limited to Europe at that time. In the US, recipe laws were primarily aimed at preventing 'economic adulteration, by which less expensive ingredients were substituted so as to make the product inferior to that which the consumer expected to receive when purchasing a product with the name under which it was sold'.[87] In short, the US standards were not conceived as promoting trade but rather as a tool of consumer protection.

In 1985, the European Commission decided to abandon its gigantic effort to introduce universally applicable 'recipe laws' for all European-made foodstuffs, as it was impossible for states to reach unanimous agreement in the adoption of the directives. In accordance with the new strategy, EC food legislation would henceforth be limited to the harmonization of national rules justified by the need to protect public health and other consumer interests, notably consumers' need for information and the necessity to ensure fair trading and provide appropriate official controls.[88] As a part of this approach, the EC Community adopted some framework directives, the so-called 'New Approach directives',[89] dealing with essential requirements in the fields of additives, labelling foods for particular nutritional needs, hygiene and official controls in order to establish basic standards and guiding Member States in the development of more detailed rules. National food legislations, constrained by the respect of the framework directives, would have been accepted within the EC by virtue of the mutual recognition principle.

In the mid-1990s, in the wake of several food outbreaks and food scares, it became clear that the free movement of foodstuffs could no longer be the overriding principle of EC food law. Food safety was not only a consumer's concern, but also an essential condition for the proper functioning of the internal market. It was therefore necessary to figure out how to reshape this European policy.

In 1997 the Green Paper on the General Principles of Food Law in the EU (Communication 1997a)[90] aimed at launching a public debate on how the EC should best regulate the area of food law. The next step in reshaping the EC food law was the Commission's publication of the 'Communication on Consumer Health and Safety' in May 1997.[91] This text set out the action which it was taking to reinforce the manner in which it obtains and makes use of scientific advice, and operates its control and inspection services, in the interests of consumer health and food safety. At that time the DG for Consumer Protection and Health (DG SANCO) was established.[92] The Commission had in particular placed the management of all the Scientific Committees working in the field of foodstuffs and responsibility for inspection and control under the authority of this DG. In December 1999, the blueprint for a European Food Authority was sketched out.[93]

The BSE crisis, growing consumer concerns about the safety of GM foods and, lastly, the dioxin contamination outbreak in Belgium, contributed heavily to spreading awareness among citizens and institutions. In the wake of these food emergencies and consumer scares, the European Commission proposed to combine the envisaged radical reform of the EC food regulatory framework with an innovative institutional reform by publishing the White Paper on Food Safety on 12 January 2000.[94] The guiding and somewhat 'revolutionary' principle of the White Paper was that food safety policy must be based on a 'comprehensive, integrated approach' throughout the food chain; across all food sectors; between the Member States; at the EC external frontier and within the EC; in international and EC decision-making for and at all stages of the policy-making process. The assumption was that a comprehensive, integrated, approach would lead to a more coherent, effective and dynamic food policy. It took more than 2 years for the Commission to transform the White Paper into a proposal for a regulation 'laying down the general principles of food law, establishing the European Food Authority, and providing for urgency measures in matters of food safety'. This proposal, published in March 2001, contained all the main features originally sketched out by the White Paper and was subsequently adopted, with few amendments, as Regulation (EC) No. 178/2002 on 28 January 2002.[95]

This regulation represents the first attempt to address all aspects of food safety at EU level by laying down a comprehensive EU food policy covering horizontally all stages of production, processing and distribution of food and feed (from farm to fork), thus encompassing raw materials, intermediate products and finished food products as well as feedstuffs. Being addressed to not only to the EC institutions, but also to the Member States, the scope of this policy is unusually broad. Undoubtedly, the establishment of the much-awaited European Food Safety Authority (EFSA) is the most prominent innovation introduced by the new European food regime. However, it must be borne in mind that the Authority merely represents one of the components of an entirely new food safety strategy adopted by the EU.

In comparison with the imposing institutional and substantive framework of the US FDA, the newly established EFSA comes across as a weak authority. Both its resources and powers are far from being like those that make the FDA one of the most authoritative administrative agencies in the world. However, it must be noted from the outset that these agencies pursue different missions: while the FDA protects the public health of American citizens, by monitoring the safety and effectiveness of products entering the market (or already in use) and by enforcing the Food and Drug Act against those who are in breach of its provisions, the EFSA is a scientific advisory body charged with providing independent and objective advice on food safety issues associated with the food chain.

The universe of the EFSA differs greatly from that of the FDA in many respects. EFSA's decisions are less science-based than those adopted by the FDA (FDA risk managers are supposed to rely exclusively on scientific factors and not on any 'social factor') and EFSA derives its powers from a regulation merely laying down general principles and requirements of food law, while FDA derives power from a text such as the Food, Drug and Cosmetic Act.

Moreover, the EFSA has not only been denied both regulatory and enforcement powers, but seems to play a limited role even within its own area of competence—risk assessment. Although it provides a common standard for conducting risk analysis throughout the EC to be followed by both the EC itself and the Member States, the current regime does not empower the EFSA to impose its own scientific vision on the Member States' competent authorities in case of diverging opinions.

Unlike the US, where arguably all local food differences have been obliterated,[96] Europe still has long-standing culinary traditions symbolizing strong identity values. Thus, a claim by a domestic food authority that a certain good is safe or unsafe is likely to involve not only an assertion about science, but also the willingness of this country to bear or not to bear the level of risk considered acceptable in order to continue or reject a certain local tradition. In contrast, the assertion made at the EC level about the safety of a product to be marketed throughout the EC is both a claim about its risk component and a political claim aimed at favouring economic integration and free trade within Europe. Along these lines, conflicts about food safety within the European context inevitably involve a tension between a European (universal) and a national (local) vision both of safety and of the sociocultural perception of a particular food.[97]

The main institutional and substantive differences between the EFSA and the FDA may be ultimately understood not only as a result of different systems of governance and regulatory environments, but also as a reflection of entirely different societal contexts. In recent years, Europeans and Americans have developed conflicting perceptions of risks and, inevitably, different cultural norms regarding food safety. This seems to be due to the complex interplay of inherent cultural models that we, as humans, use to interpret the environment and the world around us. In other words, the public perceives the risk within its own cultural model.

3.6.1   European Food Waste Policy and Legislation

In the framework of food legislation, it is only in the last few decades that Europe has progressively established a consistent political and regulatory framework on waste and food waste management. The aim of the legislation is to provide decision-makers with directions and targets, and this has led to the growth of a European waste industry that is leading the way and setting the pace worldwide.

The management of food waste involves several policy areas, including sustainable resource management, climate change, energy, biodiversity, habitat protection, agriculture and soil protection. European legislation and the waste policies of individual countries drive the separate collection of food waste. Austria, Belgium, Germany, Switzerland, Italy, Norway, Luxembourg, the Netherlands and Sweden have waste policies that are implemented nationally. These countries have pioneered separate food waste collection systems. The UK has only recently begun to tap into the experience of these other countries.

In 1989, the US report 'Facing America's Trash, What next for Municipal Solid Waste?'[98] identified that European countries were already starting to recycle food waste. There were 71 source separation projects operating in West Germany, serving 430 000 households and collecting an estimated 91 kg per person of organic waste a year. Austria, Switzerland, Sweden, Denmark and Italy were also becoming heavily involved, with France, Belgium and the UK seen as lagging behind.

The strategy on waste prevention and recycling was one of the seven thematic strategies set out in the European Sixth Action Programme for the Environment adopted in 2002.[99] In 2006, a paper reporting on the status of organic waste recycling in the EU suggested that, with the exception of Greece, large parts of Spain, Portugal, France and Ireland, all the old European countries had started to collect organic waste separately to recycle it.[100] Across Europe 1800 commercial composting sites were identified, treating 17 Mt of organic waste. Treatment of food waste by anaerobic digestion was also reported, with an estimated total capacity of 3.5 Mt. In Germany and Austria the trend to collect food waste and treat it by anaerobic digestion was seen to be increasing, driven by subsidies for renewable energy.

In recent year a sort of route map has been established for waste prevention and a few directives have been issued:

Waste Framework Directive. The Waste Framework Directive[101] or Directive 2006/12/EC of the European Parliament and of the Council of 5 April 2006 aims to protect human health and the environment against harmful effects caused by the collection, transportation, treatment, storage and disposal of waste (in this case general waste). It also establishes the principles of the waste hierarchy: reduce, re-use, recycle—in order to minimize disposal (Figure 3.2). It also defined 'material recovery targets' to drive implementation of recycling strategies. The Directive has a number of key features:

- It establishes the EU's first ever general recycling targets, incorporating household waste (50% by 2020) and non-hazardous construction and demolition waste (70% by 2020) for the first time.

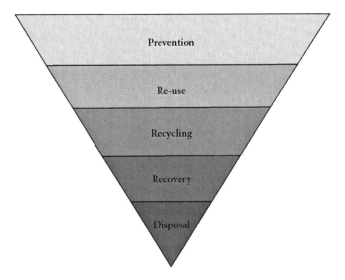

Figure 3.2   The waste hierarchy as defined in the Directive, with priority option at the top of the hierarchy and least preferred option at the bottom. Source: http://eurlex.europa.eu/LexUriServ/LexUriServ.do?uri= OJ:L:2008:312:0003:0030:EN:PDF

- It establishes a definition of 'by-products' that allows some materials currently defined as waste to become non-wastes and be removed from waste regulation, as well as the definition of minimum requirements for 'end-of-waste' criteria (*i.e.* the criteria for a waste to become a product or secondary raw material).
- It clarifies the definition of recycling to exclude energy recovery and reprocessing into fuels or backfilling materials and establishes waste prevention objectives for the first time (although firm targets were not set) to include the production of national waste prevention programmes by the end of 2013 to be articulated with mandatory waste management plans.
- Most importantly, it establishes waste prevention and decoupling objectives for 2020 and supports the ability for Member States to implement stronger measures to reduce waste.

Figure 3.2 reflects what should happen in terms of the allocation of resources and priority to the waste hierarchy. At the moment of course, much of the activity and funding mechanisms reflect the opposite, with a large financial priority still given to disposal and recovery operations.

New Waste Framework Directive. The new Waste Framework Directive[102] was adopted by the European Parliament on 17 June 2008. It provides the most significant revision of EU waste management policies since their inception in 1975 and replaces three existing directives: the Waste Framework Directive described above, the Hazardous Waste Directive and the Waste Oils Directive. It importantly requires Member States to produce and implement mandatory waste management plans that are also required to be properly evaluated.

The revised Directive sets new recycling targets to be achieved by Member States by 2020, strengthens provisions on waste prevention through an obligation for Member States to develop national waste prevention programs and a commitment from the EC to report on prevention and set waste prevention objectives. It also sets a clear, five-step hierarchy of waste management options in which prevention is the preferred option and it also draws a line between waste and by-products and defines end-of-waste criteria.

Furthermore, there is a clear strategy towards the separate collection and treatment of bio-waste. Article 22 on 'Bio-waste' says: 'Member States shall take measures to encourage the separate collection of bio-waste with a view to the composting and digestion of bio-waste, the treatment of bio-waste in a way that fulfils a high level of environmental, the use of environmentally safe materials produced from bio-waste.'

Member States had time until 12 December 2010 to transpose the provisions of the Directive into their national legislation and administrative arrangements.

Packaging Directive. The Packaging Directive[103] defined targets and therefore established drivers for the growth of the recycling industry. This Directive aims to harmonize national measures in order to prevent or reduce the impact of packaging and packaging waste on the environment and to ensure the functioning of the internal market. It contains provisions on the prevention of packaging waste, on the re-use of packaging and on the recovery and recycling of packaging waste.

In 2004, the Directive was reviewed to provide criteria clarifying the definition of the term 'packaging' and increase the targets for recovery and recycling of packaging waste. In 2005, the Directive was revised again to allow new Member States transitional periods for attaining the recovery and recycling targets.

Landfill Directive.   The Council Directive 1999/31/EC of 26 April 1999, also known as the Landfill Directive,[104] sets as a policy target the staggered reduction of biodegradable municipal waste (BMW) going to landfill. The Landfill Directive places an absolute target on the tonnage of BMW that can be landfilled by 2006, 2009 and 2016 by linking the quantity permitted to the quantity produced in 1995. Thus the Directive obliges Member States to reduce the amount of biodegradable waste in landfills by 65% by 2016 compared to 1995 levels.

This means, for instance, that if BMW production doubles between 1995 and 2016, only 17.5% of BMW produced in 2016 can be landfilled. As of 2006, Member States are restricted to landfilling a maximum of 75% of the total amount by weight of BMW produced in 1995. This target becomes 50% in 2009 and 35% in 2016. However, the Landfill Directive does not submit countries to binding specifications on methods for disposal of BMW not sent to landfills, a situation which has led most Member States to opt for incineration.

The Landfill Directive stipulates that landfilled bio-waste should be progressively reduced. Bio-waste must not go to landfill, but we are not told which of the alternative options—from composting, AD, mechanical-biological treatment, incineration—we ought to use.

Local strategies and practice regarding treatment of bio-waste need mutual consistency of plans, investments and operational behaviour. The key point is that, without a common EU perspective, choices are made by different authorities (central governments, local authorities, contractors, experts) who often go in different, or even opposite, directions. This is therefore not a safe environment for the private and the public sector to plan and deploy investments.

Bio-waste Directive.   A Bio-waste Directive[105] was planned as early as 1999, and ever since has been at the core of the waste debate at EU level. After two drafts (issued in 1999 and 2000) a 'working document' was discussed in 2003. Albeit with differing goals, those documents investigated opportunities and the potential impact of a Directive. Then at a certain point, the issue was merged with the (then ongoing) definition of the soil strategy, which equally required an EU policy to promote recovery of clean organic sources included in bio-waste. More recently, a Green Paper and an Extended Impact Assessment have been issued and discussed among stakeholders.

In December 2008, the Commission published a Green Paper on bio-waste management in the EU and launched a consultation process to gather opinions on whether a specific stand-alone EU Bio-waste Directive was needed. The purpose of the Green Paper was to explore options for the further development of the management of bio-waste by reviewing the current situation of bio-waste management in the EU. In discussing the issue over time, the key players and associations in the waste sector have realized that their views regarding bio-waste overlap to a remarkable extent. They have formed an organization called the 'Bio-waste Alliance', which has produced a few position papers and statements calling on the European Commission to continue with its plans for a Bio-waste Directive. According to their request, it is important that a Directive does not include only regulatory aspects such as conditions for compost application and so on, but also, and more importantly, 'drivers' and targets, similar to those included in the Packaging Directive, which will trigger and consolidate establishment of bio-waste-specific initiatives and secure long-term safety of plans and investments by the public sector and the waste industry.

The Commission Communication of 21 December 2005, 'Taking Sustainable Use of Resources Forward: A Thematic Strategy on the Prevention and Recycling of Waste',[106] sets out guidelines for EU action and describes the ways in which waste management can be improved. The aim of the strategy is to reduce the negative impact on the environment that is caused by waste throughout its lifespan, from production to disposal, *via* recycling. This approach means that every item of waste is seen not only as a source of pollution to be reduced, but also as a potential resource to be exploited.

In the absence of a a common EU perspective, in many Member States, the difficulty of chasing food waste management strategies caused by rapid shifts in local policy has had a negative impact on the willingness to invest in new/innovative strategies and treatment facilities. Setting targets is key to establishment of long-term confidence by decision-makers, waste planners and the recycling industry, since it establishes a common policy framework and objective, which enable them to work more effectively.

An overview of the current state of play among national composting strategies clearly demonstrates that well-developed composting systems have been completed or are being completed only in countries where drivers have been set (*e.g.* The Netherlands, Austria, Germany); others (*e.g.* Sweden, Italy and the UK) showing new developments, have recently established recycling and composting targets.

It should be noted that policies to divert food waste from landfill will not tackle the bigger issue of food waste generation. The impact of waste policy on food waste, such as the waste prevention specifications of the revised Waste Framework Directive, the Landfill Directive, and the Communication on Future Steps in Biowaste Management in the European Union, is considered to be neutral in terms of the absolute amounts of waste generated.

Waste policy does, however, have a considerable impact on the treatment of food waste once it has been generated. It has been forecast that by 2020 the amount of food waste sent to landfill will decrease from about 40.5 Mt to 4.0 Mt in compliance with policy.[107] This leaves an estimated 122 Mt of food waste across the EU-27 by 2020 still to be managed by other residual treatment technologies. Without successful long-term pan-EU waste prevention activities achieving notable behaviour change in the way people buy and use food, the treatment capacity required to handle food waste will need to more than double. The challenge this poses for raising capital, securing permission to build and planning (or extending existing facilities) will be considerable.

Discussions on how to improve EU food policies and food waste strategies have been ongoing for decades. The two EU discourses— one of economic efficiency and high technological innovation (competitiveness) and the other of environmental and social progress (sustainable development)—are now in some tension. At national level, most EU countries are implementing strategies, like for example:

Sweden. Sweden was the first country to release guidelines for 'environmentally effective food choices' (2009)[108] to educate consumers on the environmental and health impacts of various foods, and one of the first countries to set targets regarding food waste. The guidelines—based on scientific reports from the Swedish University of Agricultural Sciences (SLU) and the Institute for Food and Biotechnology (SIK)—consider the environmental and health impacts of meat (beef, lamb, pork and chicken), fish and shellfish, fruit, vegetables and legumes, potatoes, cereals and rice, fats, and water on health, climate, and the environment. Sweden's targets regarding food waste establishes that 35% of food waste from households, restaurants and stores should be recycled by biological methods in 2010 and waste from food industries should be recycled in 2010.

Recently a Swedish company has helped the South Korean city of Ulsan to convert wastewater generated from processing food waste into biogas. When garbage is processed into compost, it creates wastewater.[109]

Ireland.   Ireland launched the Food Waste Regulations on 1 July 2010.[110] The Regulations require all major sources of food waste to be placed into a dedicated bin and not mixed with other waste. A brown bin collection service must be used so that the collected food waste is subsequently recycled by composting or by other approved recycling process. Alternatively, waste producers can bring the food waste directly to a food waste recycling plant; or the food waste can be treated by composting it on the premises where it is generated. The food waste regulations apply to businesses which produce more than 50 kg of food waste per week and is therefore applicable to restaurants, bistros, cafés, public houses, canteens, industrial or office buildings where food waste is produced, hotels, guest houses and hostels, shops and supermarkets selling food, deli counters, healthcare sectors (hospitals, nursing homes), stations, airports, ports, harbours and marinas, and events/exhibitions where food waste is produced.

Germany.   In Germany the Council on Sustainable Development has developed advice for shoppers and published a booklet on this topic. The booklet[111] differentiates between frequent shopping decisions and larger, rarer purchases and decisions and affirms that sustainable consumption is an alternative to modes of consumption that fail to take such matters as resource use and the social impact of products into account.

The Netherlands. The Netherlands has produced a policy document entitled 'Towards Sustainable Food Production & consumption' (June 2008)[112] whose aim is to stimulate sustainable innovations in the Dutch agrifood complex, and enable and entice Dutch consumers to buy sustainable (and healthy) food. It also covers the national challenges and the contribution the Netherlands can make to sustainability. The Dutch approach focuses on more efficient use of space, water, energy and transport, more sustainable use of raw materials to improve soil fertility and biodiversity, countering food wastage and the loss of protein sources, and investing in innovation for better and more sustainable agriculture and food production.

A food system that respects the carrying capacity and the vitality of the global ecosystem is required alongside an innovative and dynamic agrifood chain (suppliers, producers, processing industry, transport, trade, sales).

United Kingdom. The UK has issued many policy documents such as Food 2030 (January 2010) [113] with the goal of developing and supporting sustainable supply chains for sustainable diets. In this document the UK Government sets out how the food system should look in 2030, and how it is possible to get there. Producing more food in ways that protect and enhance the natural environment, investing in the skills and the knowledge that will help the industry prosper, improving labelling so that consumers can make informed choices, reducing greenhouse gas emissions from the food chain and cutting food waste are among the most important priorities.

The UK Government published a strategy proposal in December 1995 for the sustainable management of waste in England and Wales, entitled 'Making Waste Work' (Department of Environment and Welsh Office, 1995). This document set out the government's waste management policies by building upon the ideas set out in the government's Sustainable Development Strategy (published in January 1994). The proposed strategy adopted three key objectives:

- to reduce the amount of waste produced
- to make the best use of what waste is produced

- to choose waste management practices that minimize the risk of immediate and future environmental pollution and harm to human health (Department of Environment and Welsh Office, 1995).

This was followed in June 1998 by a document called 'Less Waste: More Value'.[114] The aim was to establish a public view on the management of waste. Responses to 'Less Waste: More Value' were used to inform and direct the development of a draft Waste Strategy for England and Wales. During 2000 and 2001, a finalized waste strategy was published for England and Wales, as well as separate strategies for Scotland and Northern Ireland. Initiatives and strategies vary from region to region. England has elaborated a Waste Strategy Report 2007[115] whose key objectives are to:

- decouple waste growth (in all sectors) from economic growth and put more emphasis on waste prevention and re-use
- meet and exceed the Landfill Directive diversion targets for BMW in 2010, 2013 and 2020
- increase diversion from landfill of non-municipal waste and secure better integration of treatment for municipal and non-municipal waste
- secure the investment in infrastructure needed to divert waste from landfill and for the management of hazardous waste
- get the most environmental benefit from that investment, through increased recycling of resources and recovery of energy from residual waste using a mix of technologies.

The UK Government is therefore simplifying the regulatory system, making it more proportionate and risk based, through waste protocols that clarify when waste ceases to be waste (and so not subject to regulation); reforms of the permit and exemption systems and the controls on handling, transfer and transport of waste (with cost savings to business and regulators, *e.g.* at least £90 million on permits); and better and earlier communication with all stakeholders.

3.7 FOOD WASTE LEGISLATION IN OTHER COUNTRIES

3.7.1 Japan

Under Japan's Basic Law for Establishing a Recycling-Based Society[116] which entered into force in January 2003, the Food Recycling Law took effect in June of the same year. This law's aims are to reduce the amount of food waste generated by food manufacturers and restaurants, and to promote the re-use of food waste such as by turning it into livestock feed and compost. Pressed into action by this law, an increasing number of food manufacturers and restaurants are working to use food waste as compost.

One of those initiatives is by the Hotel New Otani in Tokyo.[117] This hotel previously had a contractor deal for its waste, but today it turns not only food waste such as banqueting leftovers but also used flowers from hotel wedding ceremonies into compost in a composting facility set up below the hotel. The compost is used by farmers contracted to grow vegetables, which are then purchased by the hotel. The hotel was able to save the costs it once paid to a waste management company, and recovered the cost of the new in-house composting facility in about 3 years. There is an increasing number of initiatives like this around the country to create a circular flow of materials between producer and consumer. For example, at several large shopping centres, food waste generated at food stalls and restaurants are thrown into a tank called a bioreactor, which then produces methane gas to run a boiler and heat water.

These efforts have led not only to a reduction of food waste but also to cost reductions. In Japan, where waste disposal sites are running out and disposal costs are increasing every year, this kind of waste reduction initiative is being pursued in earnest as a cost reduction strategy at many corporations and factories.

An interesting example is provided by the town of Aya[118] (with about 7600 residents) in Miyazaki Prefecture, where the local government collects food waste. As a collection fee, each household is charged 100 yen (about US$1) and retail stores are charged 200 yen (about US$2) per month. The food waste collected is composted and used by local farmers to grow vegetables, which are then consumed by local people. The food waste is then collected again. In this way, the nutrients are recycled locally. This initiative is beneficial not only for reducing waste but also for increasing local food self-sufficiency. In Nagai City food waste generated at households and commercial facilities is being composted, then used to grow organic produce to promote 'local production, local consumption.' Sapporo City is encouraging residents to compost using cardboard boxes.[119] The composting method requires only a cardboard box, soil conditioner purchased at a gardening store, a shovel, a thermometer and a scale. An increasing number of people are trying composting using earthworms and food waste treatment machines, some powered by electricity and some hand-powered.

3.7.2 India

India is the third largest food producer in the world, but only 2% of the food produced in India is processed.[120]

A recent estimate from the Ministry of Food Processing says a enormous Rs 580 billion[121] worth of agricultural food items are wasted in the country every year.[122] The greatest food waste comes from the agricultural waste sector and is mainly due to lack of post-harvest infrastructure such as cold chain facilities, transportation and proper storage facilities. The Bangalore Declaration, a document released by the Karnataka Government,[123] stated that nearly 120 Mt of food grain have recently been lost to rodents across the country. The federal government is therefore called to improve the country warehouses as well as to set up cold storage facilities and the Supreme Court of India is at the moment working on laws that facilitate the donation of food to the poor who do not get a daily meal.

At municipal level excellent results have been obtained since the application of the Municipal Solid Waste (Management & Handling) Rules[124] which came into force in 2000.

Currently the population of big cities generates only 50–100 g of non-biodegradable waste per capita per day. In many cities, private groups do doorstep collection for payment. In Calcutta[125] regular municipal staff using wheelbarrows collect 80% of the total waste through house-to-house collection. In cities such as Pune and Bangalore a union of women wastepickers collects both dry waste for recycling as well as food waste, for a fee. Slums are particularly efficient: 419 slums in Mumbai have a takeaway bin system.

Hotels generally collect food waste: non-vegetable food waste goes to piggeries, and leftovers to night-shelters or orphanages. In city markets all over India food waste is collected stall-to-stall, hourly.

3.7.3 Brazil

In Brazil, which is considered to become one of the top five economies worldwide within the next five years, 39 000 t of food are thrown away every day while in the meantime, 14 million Brazilians are starving.[126]

Some laws have recently been passed to reduce food waste. A highly successful recycling programme is running in the town of Curitiba;[127] 10 000 households participate in the 'Garbage that is not Garbage' programme, receiving 2 kg of food for every 4 kg of recyclable garbage collected and delivered to the mobile units. The programme was implemented to foster the separation of organic from inorganic garbage as part of the city's environmental programme. Even admittance to the municipal open air shows requires bringing a bag for recycling rubbish. The goals for the future are to transform Curitiba into a centre of excellence in the areas of urban planning and transportation, and demonstrate the success of good city planning in developing countries.

An analysis of Brazil's recycling potential including composting found that 72.8% of waste reclamation is possible. The financial cost of incineration (even with energy recovery) is calculated to be almost three times higher than the cost of recycling. Implementation could achieve a 60% beneficial use within 5 years and solve the country's escalating waste problem.[128]

Experts from Writtle College in Essex, UK[129] have been commissioned to help Brazil tackle the issues of post-harvest food wastage as the country prepares to host the 2014 football World Cup. The first focus will be on advising the state of Minas Gerais in Southern Brazil on post-harvest technology. Specifically, advice will be provided on the use of a new perishable freight system which could support the volume transport of locally grown fruit and vegetable crops including the famed 'Palmer' mango which is regularly freighted to Europe.

3.8   ROLE OF NON-GOVERNMENTAL ORGANIZATIONS AND FOOD ACTIVISM IN SHAPING FOOD POLICIES

As mentioned earlier in the chapter, some scholars have identified public pressure as one of the main drivers of policy change in the food arena, reflecting concerns about health and the state of the environment.[130]

Globally, national and transnational civil society movements are emerging as powerful advocates for a more equitable world, demonstrating that there is broad popular support in both developed and developing countries for addressing hunger and issues like food safety and sustainability.

The number of non-governmental organizations (NGOs) and their influence on global public policy have increased dramatically, especially in the 1980s and 1990s. According to the Union of International Organizations, the number of internationally operating NGOs increased from about 13 000 in 1981 to over 47 000 by 2010.[131] National numbers are even higher: Russia has 277 000 NGOs;[132] India is estimated to have around 3.3 million.[133]

NGOs are responding to food policy issues by taking increasingly rights-based and participatory approaches. NGOs now attempt to intervene to protect small farmers from eviction, indigenous people from losing traditional lands and fishing grounds, and segments of the population from discriminatory food supply schemes. NGOs are developing the concept of nutritional rights, as opposed to the right to adequate food, putting pressure on governments to take responsibility for supplying funding for nutrition in national budgets.

For example, Inter-Action, an umbrella group of development and relief NGOs based in North America, is effectively lobbying for policy support in the US, as is the US Alliance Against Hunger; and Bread for the World, an American anti-hunger advocacy group, has found that religious communities are a core constituency on hunger issues. In Brazil, political processes initiated by civil society led to the development of the Zero Hunger Program. In addition, grassroots groups in Africa have used the Poverty Reduction Strategy Paper processes to reduce corruption, strengthen democracy and improve health and education service for impoverished and hungry people.

Hence, many NGOs and civil society organizations are deeply engaged in coping with food emergencies and in providing support services to small farming communities and households, often with an emphasis on sustainable land use practices and nutrition education. Others have played a prominent role after 1996, in the consultative process on the right to food after the World Food Summit, led by the High Commissioner for Human Rights. These organizations are likely to form coalitions, taking advantage of improved networking possibilities, and to become increasingly effective forces in ensuring greater international and national commitment to addressing food issues.

In March 2010 a broad alliance of European, national and local civil society organizations concerned with the future of food and agriculture in Europe, launched a 'European food declaration'[134] whose aim was to mobilize European citizens and authorities to reshape the Common Agricultural Policy. Nearly 200 organizations from 24 Member States have so far signed the declaration and believe a strong message is needed, not only for EU policy-makers, but also for national policy-makers. The declaration calls for a complete overhaul of the current system and outlines the policy objectives of a new Common Agriculture and Food Policy for the future.

Furthermore, on 29 November 2010 the same coalition of European social movements, farmers' organizations and NGOs demanded access to quality affordable food and fair market prices for farmers: employment, payments for small farms, and the maintenance of payments coupled to production to sustain farms in disadvantaged rural areas.[135]

The coalition believes that a real reform is needed to promote environmental and socially responsible forms of farming, strengthening food production to feed people in their own regions, instead of an export-oriented and import-dependent model, as well as promoting sustainable family farms instead of large industrialized units. This reform asks for regulatory tools such as supply management, intervention and public storage in case of conjunctural sectorial crisis and border regulation to avoid low cost imports.

In the UK the democratic experimentalism brought about by civil society is bubbling. For example, the UK Food Group (UKFG)[136]—the leading UK network for NGOs working on global food and agriculture issues—is advocating for sustainable and equitable food security policies and aims at strengthening the capacity of civil society to contribute effectively to international consultations on food security. UKFG represents more than 30 development, farming, consumer and environment organizations, drawn together by a common concern for food security.

WWF-UK[137] has launched a One Planet Food Strategy 2009–2012 which advocates for a reduction in greenhouse gas (GHG) emissions from the food economy by 70% in 2050 and at eliminating unsustainable impacts on water, and change trading patterns and governance structures. One Planet Food recognizes the need to address both demand-side (food consumption) and supply-side (food production) issues within the food supply chain in order to achieve the ambitious targets for change. It takes a collaborative approach and aims at supporting the civil society, food businesses and governments.

Compassion in World Farming,[138] another UK-based NGO founded in 1967 by a British farmer who became horrified by the development of modern, intensive factory farming, campaigns peacefully to end all cruel factory farming practices and has recently launched an 'eat less meat' campaign.

Common to all these hybrid movements is the idea that policies can be changed for the better but the process is long and requires imagination, focused efforts and coalitions.[139] Public policy is regarded as an outcome of political competition among organized groups. Because of their capacity to generate publicity, the NGOs have become powerful voices within governments and effective researchers in how food is produced, processed, distributed and marketed. They are challenging governments to reshape food governance.

One outcome of NGO activities has been the emergence of counter-groups like the fast-food industry's Center for Consumers Freedom or the American Council for Fitness and Nutrition, a coalition set up in 2002 by leading food and beverage groups to counter charges according to which industry is responsible for obesity and health problems.

NGOs are a positive force in shaping food culture and policies but because they are neither governments nor commerce, they lack ultimate power. Their strengths lie in the ability to build political pressure. NGOs have contributed to a significant growth in the moral, ethical and environmental aspects of food in public culture.

Alongside NGOs, what represents the real novelty in the food policy scenario is the creation of independent food movements and pressure groups such as animals rights activists, vegetarians, vegans and many others.[140] Although many of these movements originated in the early 1980s *via* small, informal networks and organizations (such as Parents for Safe Food and the Nestlé Boycott in the UK), it is mainly in recent times and thanks to the widespread use of new communication technologies and social platforms like Facebook, Twitter and blogs, that they have become more and more popular.

Among the many relevant blogs, the following are considered as the main reference points for food activists:

- **Wasted Food:**[141] the blog of journalist and author Jonathan Bloom, who tracks legislation, reports, and news about food waste, and all in a very readable, accessible, actionable way.
- **Food Not Bombs:**[142] the blog of a movement to fight hunger, poverty and homelessness. Food Not Bombs believes 'food is a right, not a privilege' and is now a collective with chapters on every continent.
- **La *Via* Campesina:**[143] This international organization supports family farm-based production and food sovereignty: a sustainable system where the people that produce food determine their own agricultural policies.
- **Nourishing the Planet:**[144] The Worldwatch Institute publishes *Nourishing the Planet* as part of its research project examining sustainable agricultural practices and policies. The project takes a holistic approach, looking at climate change, farming and the food industry. The group publishes the State of the World report, with recommendations for poverty reduction.
- **Demonstration Nation:**[145] The tag line is 'a blog about activism', but not just any kind of activism. Taylor Leake's blog covers food policy, workers' rights, and sustainable farming. Leake writes about labour issues and the food industry for Change.org's Sustainable Food blog and *The Huffington Post*.

In 2009 in San Francisco, a handful of these and other bloggers took the week-long Hunger Challenge, which asked them to cook and eat on $4 a day (the average amount food stamp recipients spend).

All these food movements advocate for local, seasonal, healthy and sustainable food as opposed to industrialized production and mass marketing. Most of them ask people to start a garden or join a community garden group, sign petitions, meet and develop relationships with local farmers and eat seasonal food. Food activists like Alice Waters, the famous owner of the organic restaurant Chez Panisse in Berkeley, opinion makers like Carlo Petrini, founder of the Slow Food Movement,[146] actors (Meryl Streep), writers (Michael Pollan) and cooks (Jamie Oliver) are deeply committed to change the food scenario and make an impact on the public opinion.

Recently Sandor Katz's book *Inside America's Underground Food Movements* has illustrated how people are coming together to produce and obtain good food, 'real' food in the US. The book explores bread clubs, CSAs, Food Not Bombs, Slow Food, cowshares, small farms, foragers, freeganism—the collection and consumption of food which is wasted by supermarkets—and all levels of food lovers that are blazing a trail in the realm of food activism.

Movies have also been made on the controversial topic of food consumption patterns. *Food Inc*, *Taste the Waste*, *No Impact Man*, *Our Daily Bread*—all of them highlight the necessity of drastic measures to break the loop, to shift people's awareness and behaviour.

Blogs, movies, food movements, NGOs are all trying to shake the general apathy consumers are used to; they want to demonstrate that community participation can hold a number of benefits, including the generation of greater support for and sustainability of local actions and fairness and effectiveness in the setting of the food policy agenda.

REFERENCES

1. S. Henson and J. Caswell, *Food Safety Regulation*, *An Overview of Contemporary Issues*, Food Policy, 1999, p. 589.
2. S. Maxwell and R. Slater, *Food Policy Old and New*, Development Policy Review 21, 2003 available at http://www. odi.org.uk/resources/download/1238.pdf
3. http://www.food.gov.uk/foodindustry/regulation/europeleg/
4. The Codex Alimentarius Commission (Codex) was established in 1962 by two United Nations Organizations, the

Food and Agriculture Organization (FAO) and the World
Health Organization (WHO). Codex is the major international organization for encouraging fair international trade in
food and protecting the health and economic interests of
consumers. Through adoption of food standards, codes of
practice, and other guidelines developed by its committees,
and by promoting their adoption and implementation
by governments, Codex seeks to ensure that the world's
food supply is sound, wholesome, free from adulteration,
and correctly labelled. It is available at www.
codexalimentarius.net

5. United Nations, *Rome Declaration on World Food Security*,
   1996, available at http://www.fao.org/docrep/003/w3613e/
   w3613e00.HTM
6. World Health Organization, *WHO Global Strategy for Food
   Safety: Safer Food for Better Health*, 2002 available at http://
   www.who.int/foodsafety/publications/general/en/strategy_en.
   pdf
7. http://www.eoearth.org/article/World_Summit_on_Sustain
   able_Development_(WSSD),_Johannesburg,_South_Africa
8. http://www.unep.fr/scp/marrakech/10yfp.htm
9. T. Lang, *Food Industrialization and Food Power: Implications
   for Food Governance*, Development Policy Review 21 pp. 555–
   568, 2003.
10. www.ukfg.org.uk
11. www.foodfirst.org
12. www.food.gov.uk
13. S. Ambler-Edwards, K. Bailey, A. Kiff, T. Lang, R. Lee, T.
    Marsden, D. Simons and H. Tibbs, *Food Futures: Rethinking
    UK Strategy*, Chatham House Report, Chatham House,
    2009.
14. G. Tansey and T. Worsley, The Food System, Earthscan
    Publications Ltd, London (1995) in J. D. Gussow, Food
    Policy, vol. 21, issue 3, 1996, pp. 340–342.
15. T. Lang, *Food Security, Peak Oil & Climate Change: the
    Policy Context*, Talk to 'Food Security & Peak Oil' meeting,
    All Party Parliamentary Group on Peak Oil, House of
    Commons, Portcullis House, Thatcher Room, Westminster,
    March 25 2008.

16. D. Barling and T. Lang, *The Politics of UK Food Policy*, in Politics of Food, Political Quarterly Political Quarterly, 74 (1): 4–7, 2003.
17. http://www.wto.org/english/thewto_e/minist_e/min05_e/min05_e.htm
18. Grida Publications, *Rapid Response Assessment, the Environmental Crisis- From Supply to Food Security*, 2010 available at http://www.grida.no/publications/rr/food-crisis/page/3570.aspx
19. http://www.ifpri.org/sites/default/files/publications/bp002.pdf
20. J. von Braun and R. Meinzen-Dick, *Land Grabbing by Foreign Investors in Developing Countries: Risks and Opportunities*, IFPRI Policy Brief 13, April 2009.
21. http://www.new-ag.info/pov/views.php?a=783
22. J. Vapnek and M. Spreij, *Perspectives and Guidelines on Food Legislation, with a new Model Food Law*, FAO Legislative Study 87, available at http://www.fao.org/legal/legstud/ls87/ls87e.pdf
23. http://www.wto.org/english/thewto_e/whatis_e/tif_e/fact5_e.htm
24. http://www.worldtradelaw.net/uragreements/tbtagreement.pdf
25. http://www.worldtradelaw.net/tokyoround
26. http://www.codexalimentarius.net
27. http://www.caricom.org
28. http://www.crnm.org/index.php?option=com_content&view=article&id=276&Itemid=76
29. Angola, Botswana, the Democratic Republic of the Congo, Lesotho, Malawi, Mauritius, Mozambique, Namibia, Seychelles, South Africa, Swaziland, Tanzania, Zambia and Zimbabwe.
30. http://www.sadc.int
31. http://www.mercosur.org.uy
32. J. Vapnek and M. Spreij, *Perspectives and Guidelines on Food Legislation, with a new Model Food Law*, FAO Legislative Study 87, available at http://www.fao.org/legal/legstud/ls87/ls87e.pdf
33. M. Pollan, *Farm Policy & Agricultural Subsidies, A Stale Food Fight*, The New York Times, November 29, 2010.

34. M. Pollan, *The Food Issue, the Farmer in Chief*, New York Times, October 9, 2008 available at http://www.nytimes.com/2008/10/12/magazine/12policy-t.html

35. L. Tweeten, *The Economics of Global Food Security*, Ohio State University, Columbus, Ohio available at http://aede.osu.edu/programs/anderson/papers_old/EconomicsGlobalFoodSecurity.pdf

36. http://2020conference.ifpri.info/files/2010/12/20110212_parallel4E3_Johnson_Patrick_note.pdf

37. D. Imhoff, *Family Farms to Mega-Farms*, Foodfight, The Citizen's Guide To A Food And Farm Bill. Watershed Media, 2007.

38. USDA report, *The Evolving Food and Agriculture System* available at http://www.usda.gov/news/pubs/farmpolicy01/fpindex.htm

39. http://www.fns.usda.gov/wic/

40. http://www.usda.gov/farmbill

41. D. Imhoff, 2007.

42. D. Imhoff, 2007.

43. D. Imhoff, 2007.

44. U.S. Department of Agriculture *Food & Nutrition Service Homepage* available at http://www.fns.usda.gov/fns

45. K. Dahlberg, *Food Policy Councils: The Experience of Five Cities and One County* in Strategies, Policy Approaches, and Resources for Local Food System Planning and Organizing: A Resource Guide Prepared by the Local Food System Project Team, January 2002.

46. T. Philpott, *The 2007 Farm—and Food—Bill. Food First Backgrounder*, Institute for Food and Development Policy. 12:3, 2006.

47. N. Potter and J. H. Hotchkiss, *Food Science*, Aspen Publication, 1998.

48. http://michaelpollan.com/tag/farmers-markets

49. http:/www.u-s-history.com/pages/h917.html

50. http://www.fsis.usda.gov/factsheets/Additives_in_Meat_&_Poultry_Products/index.asp

51. http:// www.jstor.org/stable/30000345

52. http://www.fda.gov

53. http://www.usda.gov

54. http://www.fda.gov/Food/default.htm

55. http://www.fsis.usda.gov
56. http://www.ftc.gov
57. http://www.fda.gov/regulatoryinformation/legislation/federal fooddrugandcosmeticactfdcact/default.htm
58. http://www.fsis.usda.gov/regulations_&_Policies/Meat_Poultry_ Egg_Inspection_Directory/index.asp
59. http://www.nadreview.org
60. M. Law, *History of Food and Drug Regulation in the United States*, EH.Net Encyclopedia, edited by Robert Whaples, October 11, 2004, available at http://eh.net/encyclopedia/ article/Law.Food.and.Drug.Regulation
61. http://www.foodsecurity.org/FPC
62. A. Martin, *Into the Trash It Goes—A family of four's monthly share of American food waste*, May 18, 2008, the New York Times available at http://www.nyccah.org/files/ NYT%20Food%20Waste.pdf
63. http://www.unitedfoodbank.org/images/uploads/Good_ Samaritan_Act.pdf
64. M. Nord, M. Andrews, and S. Carlson, *Household Food Security in the United States*, United States Department of Agriculture, 2006 available at http://www.ers.usda.gov/pub- lications/err49
65. http://www.epa.gov/regulations/laws/rcra.html
66. http://www.zerowasteneo.org/group/foodwastecomposting stakeholders
67. http://compostingcouncil.org/?events=michigan-compost- operator-training
68. http://www.epa.gov/osw/inforesources/news/2007news/06- vermi.htm
69. http://www.eurekarecycling.org
70. http://www.epa.gov/osw/inforesources/news/2007news/02-ho- chunk.htm
71. J. W. Levis, *et al.*, *Assessment of the State of Food Waste Treatment in the United States and Canada,* Waste Manage- ment, 2010.
72. http://www.npr.org
73. http://www.zerowaste.org/
74. http://www.boston.com/news/education/higher/articles/2007/ 03/11/a_world_without_waste/

75. M. Hennis, *New Transatlantic Conflicts: American and European Food Policies Compared*, Robert Schuman Center for Advanced studies, Working Paper, 2002.
76. M. Hennis, 2002.
77. www.listal.com/list/make-trade-fair
78. Sicco Mansholt was the fourth President of the European Commission in 1972–1973 and the European Commissioner for Agriculture from 1958 until 1972.
79. J. Kinsey, *Will Food Safety Jeopardize Food Security?*, Proceeding 25th International Conference of Agricultural Economists, Durban, South Africa, August 2003.
80. A. Amin, *Regions Unbound: Towards a New Politics of Place*, Geografiska annaler: series B, human geography, 86, 2004, pp. 33–44.
81. http://ec.europa.eu/agriculture/cap-post-2013/consultation/consultation-document_en.pdf
82. G. Rayner, D. Barling and T. Lang, *Sustainable Food Systems in Europe: Policies, Realities and Futures*, Journal of Hunger & Environmental Nutrition, 2003, p. 145.
83. http://europa.eu/rapid/pressReleasesAction.do?reference=IP/10/1527&format=HTML&aged=0&language=EN&guiLanguage=en
84. http://ec.europa.eu/environment/eussd/
85. http://eur-lex.europa.eu/budget/data/D2010_VOL4/EN/nmc-titleN123A5/index.html
86. B. M. J. Van der Meulen, *The System of Food Law in the European Union*, available at http://www.deakin.edu.au/buslaw/law/dlr/docs/vol14-iss2/vol14–2–7.pdf
87. P. B. Hutt, and R. A. Merrill, *Food and Drug Law: Cases and Materials*, 2nd ed. New York: Foundation Press, 1991
88. *A Europe-Wide Environment Policy*, Communication from the Commission to the European Council, 29 and 30 March 1985. (Communication 1985b), available at http://aei.pitt.edu/2837/
89. http://www.newapproach.org/Directives/DirectiveList.asp
90. Commission of the European Communities, *Green Paper on European Food Law*, Brussels, 30 April 1997, available at http://www.europa.eu.int/abc/doc/off/bull/en/9704/p103042.htm

91. T. Ugland, T. and F. Veggeland, *Experiments in Food Safety Policy Integration in the European Union*, JCMS: Journal of Common Market Studies, 2006, pp. 607–624.

92. http://ec.europa.eu/dgs/health_consumer/index_en.htm

93. B.M.J. Van der Meulen.

94. Commission of the European Communities, *White Paper on Food Safety*, Brussels, 12 January 2000, available at http://www.europa.eu.int/eurlex/en/com/wpr/1999/com1999_0719en01.pdf

95. Regulation (EC) No 178/2002 of the European Parliament and of the Council of 28 January 2002 available at http://eur-lex.europa.eu/LexUriServ/LexUriServ.do?uri=CELEX:32002R0178:EN:HTML

96. E. Schlosser, *Fast Food Nation*, Houghton Mifflin, 2002.

97. D. Chalmers, *European Union Law: Text and Materials*, Cambridge, 2003.

98. *Facing America's Trash: What Next for Municipal Solid Waste?* October 1989, available at http://www.fas.org/ota/reports/8915.pdf

99. Sixth Action Programme for the Environment adopted in 2002 available at http://www.euractiv.com/en/climate-environment/6th-environment-action-programme/article-117438

100. http://ec.europa.eu/environment/waste/compost/pdf/green_paper_annex.pdf

101. http://eurlex.europa.eu/LexUriServ/LexUriServ.do?uri=OJ:L:2008:312:0003:0030:EN:PDF

102. http://eur-lex.europa.eu/LexUriServ/LexUriServ.do?uri=OJ:L:2008:312

103. http://europa.eu/legislation_summaries/environment/waste_management/l21207_en.htm

104. http://ec.europa.eu/environment/waste/landfill_index.htm

105. http://www.euractiv.com/...biowaste-directive.../article-183575

106. http://ec.europa.eu/prelex/detail_dossier_real.cfm?CL=en&DosId=193720

107. European Commission (DG ENV) Directorate C—Industry, *Preparatory Study on Food Waste Across EU27*, available at http://ec.europa.eu/environment/eussd/pdf/bio_foodwaste_report.pdf

108. http://www.se2009.eu/polopoly_fs/1.24469!menu/standard/file/Inger%20Andersson.pdf

109. http://www.usnews.com/science/articles/2009/05/21/sweden-helps-skorea-convert-food-waste-into-clean-fuel.html
110. http://www.epa.ie
111. http://www.nachhaltigkeitsrat.de/en/projects/projects-of-the-council/nachhaltiger-warenkorb/
112. http://www.minlnv.nl/portal/page?_pageid=116,1640321&_dad=portal&_schema=PORTAL&p_file_id=39545
113. http://www.defra.gov.uk/foodfarm/food/strategy/
114. http://www.docstoc.com/docs/28409401/food-waste
115. http://www.defra.gov.uk/environment/waste/strategy/strategy07/documents/waste07-strategy.pdf
116. http://www.env.go.jp/recycle/low-e.html
117. http://www.japanfs.org/en_/newsletter/200401–1.html
118. http://www.japanfs.org/en_/newsletter/200401–1.html
119. http://www.farmcentre.com/news/todaysstory/Article.aspx?id=85c303a1-b360–4012-b5e0-a44c87772718
120. http://www.india-opportunities.es/foodprocessing.php?lang
121. Equal to more than 9 billion euro http://news.bbc.co.uk/2/hi/south_asia/1695954.stm
122. http://wn.com/Ministry_of_Food_Processing_Industries
123. http://www.unido.org/fileadmin/import/41650_DynamicCity_bangalore_declaration.pdf
124. http://envfor.nic.in/legis/hsm/mswmhr.html
125. http://www.almitrapatel.com/docs/030.ppt
126. http://www.psfk.com/2009/02/ads-targeting-food-waste-in-brazil.html
127. http://www.psfk.com/ads-targeting-food-waste-in-brazil.html
128. http://www.greenpeace.org/international/en/campaigns/toxics/incineration/alternatives-to-incineration/
129. http://www.writtle.ac.uk/pge_PressRelease.cfm?ID=723
130. T. Lang, *Food Industrialization and Food Power: Implications for Food Governance*, Development Policy Review 2, 2003, p. 558.
131. S. Charnovitz, *Two Centuries of Participation: NGOs and International Governance*, in Michigan Journal of International Law, Winter 1997.
132. http://www.chicagotribune.com/news/nationworld/chi-russia-civil_rodriguezmay07,0,3849939.story
133. http://Oneworld.net
134. http://www.europeanfooddeclaration.org

135. http://www.europeanfooddeclaration.org/who-are-we
136. http://www.ukfg.org.uk/
137. http://www.wwf.org.uk
138. http://www.ciwf.org.uk/about_us/default.aspx
139. T. Lang, *Food wars, the Global Battles for Mouths, Minds and Markets*, Earthscan, 2004.
140. D. Maurer and J. Sobal, *Eating Agendas: Food and Nutrition as Social Problems*, Transaction Publishers, 1995.
141. http://www.wastedfood.com/
142. food-not-bombs.blogspot.com
143. http://www.viacampesina.org/en/index.php?option=com_content&view=section&layout=blog&id=7&Itemid=29
144. blogs.worldwatch.org/nourishingtheplanet/
145. www.demonstrationnation.com/.../blog-day-of-action-alert.html
146. http://www.slowfood.com/

CHAPTER 4

Food Waste Prevention and Reduction: Why It Is Imperative

'The stuff we throw away' is just the tip of a material iceberg . . . contain[ing] on average only 5 percent of the raw materials involved in the process of making and delivering it.

William McDonough and Michael Braungart,
inventors of the Cradle to Cradle Theory (2002)

4.1 FOOD WASTE PREVENTION

Prevention of the wastage of edible food within agriculture, industry, trade and gastronomy could save effort, resources and emissions but is difficult to implement: it requires on one hand assumptions and prognosis to prevent surpluses, overproduction and unprofitable prices and on the other hand accurate weather forecasts to avoid harvest and post-harvest spoilage.

Food waste prevention therefore requires a multilevel approach where agricultural, national and international appropriate policies and strategies integrate with consumers' awareness and effective technologies.

In order to reduce the amount of food waste to be disposed of, the US Environmental Protection Agency (EPA) has developed a hierarchy for food waste prevention (Figure 4.1), following the spirit of

Transforming Food Waste into a Resource
By Andrea Segrè and Silvia Gaiani
© Andrea Segrè and Silvia Gaiani, 2012
Published by the Royal Society of Chemistry, www.rsc.org

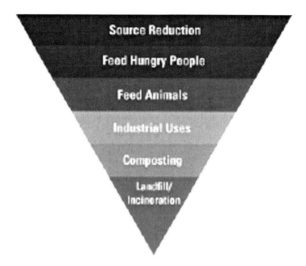

Figure 4.1 US EPA food waste recovery hierarchy. Source: www.epa.gov/epawaste/nonhaz/municipal/index.htm

the European Union (EU) waste hierarchy as presented in the 2008 Waste Framework Directive.[1] Such a hierarchy sets the guidelines for food waste avoidance, prioritizing reduction at source, and could be considered as a worldwide effective strategy. It presents a list of preference for use, re-use, recycling and waste treatment.

It should be noted that approximately one-third of all food waste is inedible, and thus options such as diversion to animal feed, industrial uses of food waste (*e.g.* cooking oils) and composting will usually be the environmentally preferable choice.

The EPA hierarchy does not differentiate between waste treatment options, but anaerobic digestion is likely to be environmentally preferable to incineration and landfilling. The food waste recovery hierarchy[2] includes the following activities, with disposal as the last and least preferred option:

- **Source reduction**: aiming to reduce the amount of food waste being generated.
- **Feeding people**: implying the donation of excess food to food banks, soup kitchens and shelters.
- **Feeding animals**: aiming to provide food scraps to farmers.
- **Industrial uses**: including the provision of fats for rendering; oil for fuel; food discards for animal feed production; or

anaerobic digestion combined with soil amendment production or composting of the residuals.
- **Composting**: implying the recycling of food scraps into a nutrient rich soil amendment.
- **Landfills**: as the ultimate option.

Source reduction actions suggest measures at different levels.

4.1.1 Household Level

Although it may sound simple, it is difficult to find an appropriate measure to implement food waste prevention at the household level. There are various reasons for this, such as age, income, time spent at home, lifestyle factors and situational factors such as smell, hunger or appetite which shape consumer behaviour. Consumption patterns, which are mainly driven by automatic actions, cannot be changed immediately by setting a specific action. It is assumed that 80% of nutrition and environmental behaviour is habitual.[3]

Recent researches estimate that about two-thirds of the food waste found in residual waste from households could theoretically be prevented. Partly used food and leftovers—which contribute up to 18% to the amount of food waste at household level—could potentially be prevented. Preparation residues, *e.g.* eggshells or vegetable peelings, represent a major part of food waste (over 30%)[4] that can hardly be avoided when fresh food is used for cooking.[2]

Effective prevention measures concerning food waste at household level include the provision of information about origin, production conditions, seasonality, the correct storage and preparation, but also the launch of awareness campaigns (like the 'Love Food, Hate Waste'[5] campaign in the UK), training programmes (Bruxelles Environnement's Anti-Waste Workshop Cooking Classes[6]) and promotional strategies, such as Tesco's 'Buy One Get One Free Later' scheme[7] that allows customers buying perishable goods to collect their free item the following week.

4.1.2 Business and Institutional Level

Initiatives at this level should focus on the hospitality industry (restaurants, hotels, catering services), schools, hospitals and similar environments. Examples of initiatives in this sector should include efforts to measure the quantity of food waste generated in their restaurants and cafeterias (the Eurest food service company in Sweden[8]), publicize results to staff and customers, explain the impacts of food waste and promote preventative actions.

4.1.3   Retail Level

Initiatives should include marketing of different grades of fresh produce at different price levels; optimizing stock rotation and ordering; and better assessment of customer demand to ensure that stock is sold in time.

4.1.4   Distribution and Wholesale Level

Initiatives should aim at developing a research programme that develop technologies to monitor food temperature data throughout the food supply chain, aiming to reduce spoilage and increase safety.

4.1.5   Manufacturing and Processing Level

Initiatives should make use of technologies to address process inefficiencies and inconsistencies, thus reducing spoilage, and research programmes to identify areas where food waste is generated for different food products, such as WRAP's Resource Maps Project[9] that supports the development of targeted waste prevention measures.

   Approaches other than reduction at the source may include the following.

4.1.6   Food Redistribution Activities

Food redistribution programmes aim to collect food that would otherwise be discarded by retailers, because it is damaged or nearing expiry, and distribute it to a variety of groups in need, including the homeless, the elderly, children and other communities in food poverty. As highlighted earlier, the quantities of edible food waste in the wholesale/retail sector are very large and present enormous opportunities to increase this sort of critical activity.

While there are many examples of food redistribution programmes, they remain predominantly on a small scale. In the UK,
for example, retailer Sainsbury's donates 400 kg of food products
to food redistribution programmes for every £1 million pounds (or
€1.1 million) in sales, similar to donation levels in the United
States.[10] For-profit enterprises that collect unsellable food from
retailers and resell it in other venues, such as discount stores, have
also proved effective in minimizing food waste.

Food redistribution programmes (*e.g.* food banks) are also very
effective at preventing food waste, and, with examples in many
EU countries (Austria, Denmark, Spain, Italy, UK) and elsewhere
(Argentina, Brazil, Canada, US), have demonstrated their repro-
ducibility and ability to adapt to local circumstances and business
opportunities.

The organizations served by food banks include community
centres, shelters, soup kitchens, food pantries, childcare centres and
senior citizen programmes. In addition there are other organizations, driven by churches and charities, *e.g.* 'Tafeln'[11] (tables) in
Germany or 'Sozialmärkte'[12] (social supermarkets) in Austria,
which manage the collection of edible food items and distribute
them to social welfare organizations or directly to people in need.
These organizations collect edible food which otherwise would be
wasted from farmers, industry and trade. The food items are sold at
a nominal price (about one-third of the usual price) to people
in need. Thus, people who have the possibility of cooking and
eating at home are supported with food without having to accept
charity.

4.1.7 Providing Food to Livestock Farmers or Animal Shelters

Converting excess human food into animal feed should be promoted, too. In many areas hog farmers have traditionally relied on
food discards to help sustain their livestock. In addition, farmers
may provide storage containers and free or low-cost pick-up
service.

4.1.8 Industrial Uses

Industrial uses of otherwise inedible food might may be successful.
An example here is represented by the 'Fish Chips' created in

Denmark, using inedible fish matter to create a marketable snack rich in omega-3 fatty acids; there are potentially many similar examples, although few have given concrete results.[13]

Apart from specific actions, a good food waste prevention strategy should implement sustainable solutions across the entire food supply chain. In developing and emerging economies, this would require market-led large-scale investment in agricultural infrastructure, technological skills and knowledge, storage, transport and distribution. Such investments have been shown to stimulate rural economies, *e.g.* the development of the Nile perch fishery in East Africa[14] has stimulated infrastructure development and considerably reduced post-harvest losses.

For long-term sustainability, development across the food supply chain in the developing world would require locally supported government policies and investment alongside any market-led private investment with reach through into developed world markets.

Conversely, the greatest potential for the reduction of food waste in the developed world lies with retailers, food services and consumers. Cultural shifts in the ways consumers value food, stimulated by education, increased awareness of the food supply chain and the impact of food waste on the environment have the potential to reduce waste production. Improved food labelling and better consumer understanding of labelling and food storage could have food waste reduction potential as well.

4.2 WASTE AUDITING: AN OPTION FOR RETAILERS?

Governments all over the world are now reflecting whether forcing retailers to declare their annual waste quantities and reveal their waste quantification methodologies could be a good way to help the reduction of food waste. Surprisingly, it seems that it would be relatively easy to audit food waste arising at large manufacturing centres. In the UK and the EU large manufacturers are already required to report on tonnages of solid waste arising in their process under a regulation known as Integrated Pollution Prevention and Control (IPPC).[15] Within this measuring mechanism, the quantification of food waste could also be introduced.

In his latest book, *Waste—Uncovering the Global Food Scandal*, Tristram Stuart suggests that a 50% food waste reduction across main food industries should be incentivized over a period of 5 years. If reduction targets are not met, then businesses should pay for the wastage of valuable food.[16]

In the first years of auditing and the adoption of food waste reduction targets, big businesses and main retailers should lead the way in demonstrating best low food waste practice. The introduction of transparent annual mandatory food waste audits across all large food industries and the introduction of fiscally driven food waste reduction targets would see a huge and lasting reduction in food waste. The scheme would not be hard to implement on a large scale, and would encourage business, based on companies being able to demonstrate to the consumer that they are committed to environmental and social responsibility.

The UK is the leading European country in terms of waste auditing: a British campaign called 'This is Rubbish'[17] was launched in 2009 and its main aim is to demand a transparent food waste auditing policy across all main food industries, and the subsequent introduction of mandatory food waste reduction targets.

4.3   FOOD WASTE DISPOSAL OPTIONS

When food waste disposal is unavoidable, there are three principal methods available: dumping, burning and minimizing.

- **Dumping** is the most common method of food waste disposal, but it may create sanitation and landfill problems. The use of landfill has always been the cheapest waste disposal option, but its cost as a waste disposal option is rising rapidly: the average cost of landfilling food waste in the UK is around £100 per tonne[18] while in the US it ranges from a low of $9.75 to a high of $37 per tonne.[19]
- **Burning** food waste has some favourable attributes such as volume reduction, immediate disposal, lower land area requirements, destruction of some hazardous materials and energy recovery. It is more widely used where the population density is high and there may not be enough room for landfills. However, incineration also has downsides: it is much more expensive than landfilling and it can release toxic substances (*e.g.* dioxins or mercury) into the air.

- **Minimizing** food waste occurs through food trades, gifts, donations and conservation during preparation and after consumption, such as reusing leftovers. Without a doubt, the best way to manage waste is to minimize, it but it is not always a viable option.

It is very difficult to assess which waste disposal system is best, as all of them affect the environment differently and depend on many factors, including the availability of municipal facilities, population density, infrastructure and climate. In areas where waste collection is a public function, food waste is usually managed by the same governmental organization that manages other waste collection. Most food waste is combined with general waste at the source. Separate collections have the advantage that food wastes can be disposed of in ways not applicable to other wastes.

From the end of the 19th century until the middle of the 20th century, many municipalities collected food waste separately. It was typically disinfected by steaming and fed to pigs, either on private farms or in municipal piggeries. Separate kerbside collection of food waste is now being revived in some areas. To keep collection costs down and raise the rate of food waste segregation, some local authorities, especially in Europe, have introduced alternate weekly collections of biodegradable waste (including *e.g.* garden waste), which enables a wider range of recyclable materials to be collected at reasonable cost, and improve their collection rates. However, this may result in a wait of up to 2 weeks before the waste is collected, so there is criticism that, particularly during hot weather, food waste rots and stinks, and attracts vermin.

Other valuable alternative to food waste disposal are recycling and composting:

- **Recycling** is when materials are collected and used again as raw materials for new products. The cycle includes collecting the recyclables, separating them, processing them, manufacturing and purchasing items made from reprocessed materials. Recycling prevents the emission of many greenhouse gases (GHGs) and water pollutants, saves energy, supplies valuable raw materials to industry, creates jobs, stimulates the development of greener technologies, conserves resources and reduces the need for new landfills and combustors. Recycling also helps reduce GHG emissions that affect global climate and is cost-effective because it reduces the quantity of wastes being sent to landfills and thereby saves landfill tipping fees.

- *Composting* is the biological process of breaking up of organic waste such as kitchen waste, manure, leaves, grass trimmings, newspaper, coffee grounds, *etc.*, into an extremely useful humus-like substance by various microorganisms including bacteria, fungi and actinomycetes in the presence of oxygen. It is a natural process which needs almost no human intervention. Households can have their own backyard compost heap in which they can put their food waste and garden trimmings so as to reduce the amount of waste that has to be collected. There are many benefits to composting food waste, including making a valuable soil product that will add biodiversity and structure to the soil to increase the health and yields of the soil, avoiding disposal fees at the landfill, helping to meet waste reduction goals and sustaining local recycling infrastructures. In addition, removing organic material from the landfill reduces the amount of methane that a landfill produces. Methane is a GHG that is 23 times more efficient than CO_2 at trapping heat in the Earth's atmosphere.[20]

Obviously, the more material is recycled or composted, the less residual waste will need collecting for subsequent treatment or disposal. Much kitchen waste also leaves the home through garbage disposal units.

Very different national policies apply to food waste management, ranging from little action in some countries to ambitious policies in others. This can lead to increased environmental impacts and can hamper or delay full utilization of advanced waste management techniques.

4.4 HOW TECHNOLOGIES MAY CONTRIBUTE TO THE REDUCTION OF FOOD WASTE

Innovative technology, in both developed and developing countries, can contribute to improve shelf life for perishable foods and semi-prepared meals, transform food waste into energy or simply track food waste.

There are many technologies that turn food waste into products by closing the cycle. For example, a start-up company launched by a group of Georgia Institute of Technology students and alumni in the US is developing software and hardware to track and monitor perishable food, from its harvesting on the farm to its arrival at the grocery store, to ensure its freshness and prevent spoilage. Fresh Test,[21] as the company is called, aims to reduce waste and spoilage in the food supply chain by incorporating technologies that monitor the status of food shipments and delivery. Customers are alerted any time a refrigerator door is left open for too long, for example, or any time the food products risk spoilage due to excessive humidity or improper pH.

Another example is provided by an American company called LeanPath Inc.[22] which has recently announced its newest software solution for food waste tracking, WasteLOGGER, which helps any restaurant, hotel, caterer or food service operation to prevent, minimize and avoid pre-consumer food waste (including over-production, spoilage, expiration and trim waste). The software runs on existing computers and does not require specialized tracking equipment. Food waste tracking is an essential aspect for any food service operation looking to become more sustainable, as it helps to strengthen efficiency and conserve resources.

LeanPath also has a partnership with Sodexo to track food waste on eight American college campuses. The pilot study focuses on kitchen (*i.e.* pre-consumer) waste, not what customers throw out, and features a tracking station where Sodexo employees enter data about what they are throwing out and why. By tracking the reason for throwing away items, Sodexo is able to correct the problem to prevent future food waste. Sodexo employees at those eight sites have dramatically reduced overproduction, spoilage, expiration and trimmings by participating in the pilot study.

An interesting case study is represented by the East Bay Municipal Utility District (EBMUD) in Oakland, California which in 2006 was given a grant by the EPA to investigate anaerobic digestion of food waste. The purpose of the study was to identify design and operating criteria for anaerobic digestion of food waste, and to compare food waste digestion to that of municipal waste-water solids digestion. Processing involves creating a slurry from the pre-sorted food waste and further reducing contaminants and food waste particle size prior to digester feeding. Waste haulers collect post-consumer food waste from local restaurants and markets and take it to EBMUD. In an anaerobic digester, the food waste is broken down by bacteria and methane is released as a by-product. EBMUD then captures the methane and uses it as a renewable source of energy to power the treatment plant. After the digestion process, the leftover material can be composted and used as a natural fertilizer.[23] This is the first sewage treatment facility in the nation to convert post-consumer food scraps to energy by anaerobic digestion.

A wide array of technologies in the field of packaging offers a range of solutions to reduce the amount of food wasted through resealable packaging, packaging of agricultural products and smart packaging that indicate when food has spoiled. DSM, a company based in the Netherlands, has developed a product called Pack-Age which allows for cheese to ripen naturally in a film with an optimum taste and texture yet without the risk of mould formation, further reducing unnecessary food waste.[24]

Changing packaging and consumer behaviour is only part of the solution. Preservation ingredients, hygienic measures and temperature-controlled transport must also be considered. With the journey from farm to fork often involving intercontinental trade, the need to prolong shelf life has become increasingly important in order to ensure that the product corresponds to what the consumer has come to expect once it reaches the store. This can be accomplished in two ways: preservation ingredients and additives.

Preservation ingredients form an increasingly important part in reducing food waste. The common functions for preservatives in food include colour retention, flavour protection, mould inhibition, spoilage retardation and general preservation. An important preservation ingredient developed by DSM is natamycin[25] (an antimicrobial agent), a 100% natural ingredient which is used for various foods, such as cheese and sausages. Natamycin is used in the food industry as a preservative as it prevents the growth of moulds and yeasts that spoil the food. Currently, in the US, this ingredient is widely used in products such as hard cheese, sausages and yoghurts. Natamycin has undergone extensive testing and the quantities applied to food products do not pose any risk to humans.

4.5 FOOD WASTE TO 2050: PROJECTIONS AND UNCERTAINTIES

In the developing world, lack of infrastructure and associated technical and managerial skills in food production and post-harvest processing have been identified as key drivers in the creation of food waste, both now and in the near future.[26]

This situation contrasts with that in developed countries where scientists forecast the majority of food waste continuing to be produced at post-consumer level, driven by the low price of food relative to disposable income, consumers' high expectations of food cosmetic standards and the increasing disconnection between consumers and how food is produced. Similarly, increasing urbanization within transitioning countries will potentially disconnect those populations from how food is grown, which is likely to further increase the generation of food waste.

Across the globe, resource and commodity limitations, in part as a result of an increasing population but also owing to impacts of climate change, are viewed as being likely to increase the economic value of food, potentially driving more efficient processes that could lead to food waste reduction.

Industrialized food supply chains will continue to develop in response to these wider challenges by the development of shared logistics (*e.g.* collaborative warehousing), identification and labelling of products (use of barcodes and tags) and better demand forecasting (Global Commerce Initiative 2008).[27] Domestic kitchen technologies (smart fridges, cookers, online meal planning and recipe resources) may make it easier for consumers to manage their food better and waste less of it.

Based on anticipated EU population growth and increasing affluence only, food waste is expected to rise to 126 Mt in 2020 from 89 Mt in 2006.[28]

In the developing world, transfer of existing technologies and the spread of good practice, allied to market-led investment, have the greatest potential to reduce food waste across the food supply chain. It is of key importance, however, that practical developments address the problems of local farmers, using indigenous knowledge where that has been shown to be sustainable. Without participation of local farmers, such knowledge transfer is unlikely to succeed.

While attempts to shift consumer behaviour may result in reduction in food waste in developed countries, changes in legislation and business behaviour towards more sustainable food production and consumption will be necessary to reduce waste from its current high levels.

An example might be through the development of closed-loop supply chain models.[29] In such models, waste of all types would be fed back into the value chain. For example, packaging waste would be re-used, and food graded as lower quality for cosmetic reasons and food that is surplus to retailer or manufacturers would be made available through alternative routes, while unavoidable food waste would be utilized as a by-product, *e.g.* in providing energy from waste using the appropriate technology.

A firm evidence base from which to assess food waste globally is lacking, with no specific information on the impact of food waste in Brazil, Russia, India and China (the BRIC countries) being a major concern, and with much of the loss estimates from developing countries collected over 30 years ago. There is a pressing need for quantitative evidence covering developing countries and the rapidly evolving BRIC countries. Without systematic evidence, the arguments over the potential for reducing global food waste as a contribution to feeding 9 billion people by 2050 will remain largely rhetorical, and measuring progress against any global reduction target impossible.

The lack of infrastructure in many developing countries and poor harvesting/growing techniques are likely to remain major elements in the generation of food waste. Less than 5% of the funding for agricultural research is allocated to post-harvest systems, and yet reduction of these losses is recognized as an important component of improved food security.[30] Worldwide, there is a need for successful introduction of culture-specific innovations and technologies across the food supply chain to reduce losses.

The related concept of market transformation has enormous potential to develop food supply chain infrastructure and reduce waste in developing and BRIC countries. Account should be taken of the impact of market transformation on the local communities to whom food may no longer be available.

Most studies show that as the proportion of income spent on food declines, food waste increases. There is clear evidence of a distribution of waste across demographic groups, with the lowest wastage rates in the generation of immediate post-war age.

However, it would be a mistake to assume that the demographic distribution will remain the same in the future. Although today's elderly generally exhibit a 'waste not want not' mentality, the elderly people of the future are likely to continue to retain the same attitudes and behaviours to food that they have today.

There are clearly fundamental factors affecting post-consumer food waste worldwide, some of which may require solutions that involve direct communication and awareness-raising among consumers of the importance of reducing food waste. Others require government interventions and the support and cooperation of the food industry itself, such as improving the clarity of food date labelling and advice on food storage, or ensuring that an appropriate range of pack or portion sizes is available that meets the needs of different households.

4.6 POLICY RECOMMENDATIONS TO REDUCE FOOD WASTE

A recent EU report on food waste financed by the DG Environment[31] identifies a series of policy options for the implementation and strengthening of existing efforts to prevent and reduce food waste at the EU level. Although the options illustrated by the report refer mainly to the EU, recommendations could be applied on a global scale as they refer mainly to food waste data reporting requirements, the clarification and standardization of current food date labels, the setting of targets for food waste prevention, the separate collection of food waste and the establishment of awareness campaigns.

The lack of reliable data on food waste is a recurring obstacle, impacting the assessment of the environmental impacts of food waste, the anticipated developments in food waste generation over time, and the setting of targeted policies for waste prevention. Legislators at international, European and national level should direct action on food waste by providing a quantitative basis for policy-making and target-setting. For example, the estimated cost of undertaking a bin characterization study at national level is expected by the Irish Environment Agency to be around €30 000.[32]

International research spanning over a decade systematically recommends the adoption of separate collection of food waste or biodegradable waste for the household and/or food service sector. The costs for separate collection vary between countries: for

Table 4.1 Estimated costs of separate collection of food waste

Costs of implementing separate food waste collection	
Cost of separate collection followed by composting	35–75 €/tonne
Cost of separate collection of bio-waste followed by anaerobic digestion	80 to 125 €/tonne
Compared with landfill and incineration	
Cost of landfill of mixed waste	55 €/tonne
Cost of incineration of mixed waste	90 €/tonne

Source: European Commission (DG ENV), Directorate C—Industry, Preparatory Study on Food Waste across the EU27, October 2010.

example, in the 27 countries of the European Union they could be estimated as in Table 4.1.

The cost of separate collection followed by composting is generally cheaper than the cost of separate collection of biowaste followed by anaerobic digestion. and landfill is cheaper than incineration. As a consequence, a legal framework (and possibly a separate directive, besides the new framework directive on waste) relating to selective collection of bio-waste and sustainable treatment would be necessary to stimulate the appropriate investment in disposals.

As well as a proper waste disposal strategy a labelling coherence policy should also become a major requirement: 'best before', 'sell by' and 'display until' dates should be unified in a unique definition so to increase public awareness of food edibility criteria, thereby reducing food waste produced due to date label confusion or perceived inedibility. Research on date labelling undertaken in the UK shows that 45–49% of consumers misunderstand the meaning of the date labels 'best before' and 'use by'. WRAP estimate that 1 Mt of food waste, or over 20% of avoidable food waste in the UK, is linked to date label confusion.[33]

To conclude, lack of awareness, lack of knowledge on methods for avoiding food waste, date label confusion and inappropriate food waste disposals could be avoided through better policies and awareness campaigns.

A concrete zero food waste strategy has to emerge from the dynamic and interaction between national and international governments and partners in the food chain. As yet there is no sustainable food system in existence, therefore ministries of agriculture, nature and food quality worldwide should provide incentives, create the necessary conditions, and enhance instruments such as knowledge and research. Clear standards and timescales must be set. Currently innovative businesses and organizations are launching many initiatives in the area of sustainability. With an extra boost from government, such initiatives should before long be regarded as the norm.

REFERENCES

1. http://www.epa.gov/epawaste/nonhaz/municipal/index.htm
2. USDA Report, *Waste Not, Want Not: Feeding the Hungry and Reducing Solid Waste Through Food Recovery*, available at http://www.epa.gov/osw/conserve/materials/organics/pubs/wast_not.pdf
3. F. Schneider, *Wasting Food—An Insistent Behaviour*, Paper presented at Urban Issue and Solutions Conference, May 11–15 2008, available at http://www.ifr.ac.uk/waste/Reports/Wasting%20Food%20-%20An%20Insistent.pdf
4. F. Schneider, 2008.
5. http://www.lovefoodhatewaste.com
6. http://www.nweurope.eu/nwefiles/file/GreenCook_project.ppt
7. http://www.dailymail.co.uk/news/article-1220842/Tesco-plans-Buy-one-free-later-deal-bid-cut-food-waste.html
8. http://www.foodoresund.com/composite-313.htm
9. http://www.wrap.org.uk/document.rm?id=10229
10. http://www.basingstokeobserver.co.uk/news/community/grub-boost-for-basingstoke-food-bank-1117
11. http://www.tafel.de
12. http://www.sozialmarkt.at
13. European Commission (DG ENV)—Directorate C—Industry, *Preparatory Study on Food Waste across EU 27*, October 2010 p. 100 available at http://ec.europa.eu/environment/eussd/pdf/bio_foodwaste_report.pdf
14. http://www.aquaticcommunity.com/mix/nileperch.php
15. http://europa.eu/legislation_summaries/environment/waste_management/l28045_en.htm
16. T. Stuart, *Waste- Uncovering the Global Food Scandal*, Penguin 2009.
17. http://www.theecologist.org/how_to_make_a_difference/recycling_and_waste/512923/campaign_fighting_food_waste_through_feasts.html

18. http://www.landfill-site.com/html/rising_costs.html

19. http://www.mackinac.org/6134

20. http://bgm.stanford.edu/pssi_food_composting_requirements

21. http://www.certifiablygreenblog.com/?p=1223

22. http://www.renewable-energy-news.info/food-waste-tracking-software-allows-sustainable-business-operations/

23. http://www.epa.gov/region9/waste/features/foodtoenergy/

24. http://www.dsm.com/en_US/cworld/public/media/downloads/publications/backgrounder_Food_Waste.pdf

25. http://www.dsm.com/le/en_US/foodspecialties/html/homepage.htm

26. United Nations *World Food Programme Report 2009*, available at http://www.wfp.org/content/annual-report-2009

27. http://www.mildpdf.com/result-2016-the-future-value-chain-global-commerce-initiative.html

28. European Commission (DG ENV)—Directorate C—Industry, *Preparatory Study on Food Waste across EU 27*, October 2010 p.100 available at http://ec.europa.eu/environment/eussd/pdf/bio_foodwaste_report.pdf

29. W. D. Solvang, Z. Deng and B. Solvang, *A Closed-loop Supply Chain Model for Managing Overall Optimization of Eco-effi-ciency*, paper presented at POMS 18th Annual Conference Dallas, Texas, U.S.A. May 4 to May 7, 2007 available at http://www.poms.org/conferences/poms2007/cdprogram/topics/full_length_papers_files/007-0582.pdf

30. C. Nellemann, *et al.*, *The Environmental Food Crisis—The Environment's Role in Averting Future Food Crises*, UNEP, GRID-Arendal, February 2009, available at http://www.ilo.org/global/What_we_do/Publications/Newreleases/lang--en/docName--WCMS_098503/index.htm

31. European Commission (DG ENV)—Directorate C—Industry, *Preparatory Study on Food Waste across the EU27*, October 2010, available at http://ec.europa.eu/environment/eussd/pdf/bio_foodwaste_report.pdf

32. http://www.epa.ie

33. WRAP, *Household Food and Drink Waste in the UK*, 2010 available at http://www.wrap.org.uk/retail_supply_chain/research_tools/research/report_household.html

CHAPTER 5

Food Recovery Programmes—Don't Let Food Waste Be Wasted

If we could recover just 5% of the food wasted each year in America, we could help feed 14 million people.

Feeding America – Hunger Relief Charity

5.1 FOOD WASTE RECOVERY PROGRAMMES: PUTTING SURPLUS FOOD TO GOOD USE

As mentioned in the previous chapters of this book, food excess, and consequently food waste, occur along the whole supply chain—in the fields, at retail level, at household level, in public places such as markets, restaurants, cafeterias, hospitals, airlines, and at special events. Reduction at source, better logistics and marketing strategies, commercial measurement and tracking software, home composting and better planning, alongside municipal separated collection, could all represent alternatives to food waste incineration and disposal.

Food waste is in fact a valuable resource that should not be lost: when food goes to waste, other resources such as water, energy, and labour hours are wasted too. Food waste should, when possible, be re-used, recycled or composted. In other words, it should

Transforming Food Waste into a Resource
By Andrea Segrè and Silvia Gaiani
© Andrea Segrè and Silvia Gaiani, 2012
Published by the Royal Society of Chemistry, www.rsc.org

be given the chance of a second life. For this reason, food waste recovery represents an important part of the food system, and in particular of a sustainable food system, as it closes the food loop.[1] As the final step in the movement of food along the supply chain, food waste can in fact become an input back into the food system and, thus converted, can become a new resource.

Food recovery takes several forms: gleaning, perishable food rescue/salvage, non-perishable food collection, and rescue of prepared food.[2]

- **Gleaning** refers to the collection of crops either from farmers' fields that have already been mechanically harvested or from fields where it is not economically profitable to harvest, due to low market prices.
- **Perishable food** (fresh fruits and vegetables, cooked/ready-made meals)—rescue/salvage is carried out from wholesale and retail outlets. In some instances food recovery applies to produce that is riper than is appropriate for transport to retail outlets.
- **Non-perishable food** (food with a relatively long shelf life) is usually collected from manufacturers, wholesalers and retailers.
- **Prepared food** rescue refers to food collected from the food service industry, *i.e.* restaurants, caterers, hotels and other commercial kitchens.

Food scraps, once recovered, can be put to beneficial use, by either being donated to hungry people (when food is wholesome, unspoiled, healthy and edible) or to animal sanctuaries, or used to feed farmers' livestock (in many areas hog farmers have traditionally relied on food discards to help sustain their livestock). Liquid fats and solid meat products can also find a second life by being converted into animal food, cosmetics, soap, and other products.

Composting is a further effective way to convert food scraps that cannot be fed to people or animals into an organic-based nutrient source for plants. Composting can be done both on-site, *i.e.* where the food scraps are generated, or off-site and can take many forms:

- **Unaerated static pile composting**: organic materials are piled and mixed with bulking material.

- **Aerated windrow pile composting**: organic materials are formed into rows or long piles and aerated either passively or mechanically.
- **In-vessel composting**: composting vessels are enclosed, temperature- and moisture-controlled systems.
- **Vermicomposting**: worms convert organic materials into a high-value compost; his method is faster than windrow or in-vessel composting and produces a high-quality compost.

According to the US Environmental Protection Agency (EPA), food waste disposal is only rarely a good option. The EPA has released a Food Waste Management Calculator,[3] which evaluates the competitiveness of alternatives to traditional food waste disposal, such as composting, source reduction, donation and recycling of fats, oils and grease (FOG). Based on inputs such as types and quantities of food waste generated by a specific organization and the availability of recovery methods, the calculator determines what is the best alternative to landfill. The calculator demonstrates that food recovery is cost-effective for many facilities and it is also environmentally and socially sound.

In the last decade, many new initiatives whose aim is to recover food and convert it into valuable end uses have spread all over the world. Most of these programmes focus on commercial and institutional sectors, but some are also tapping food recovery from the residential sector and are offering collection of source-separated food discards. One of the biggest challenges in diverting commercial food waste is overcoming the perception that segregating food waste is extra work and a nuisance. Through outreach and technical assistance, operational food recovery programmes can easily overcome this obstacle.

Currently in Europe there are hundreds of initiatives/projects aiming at preventing and reducing food waste[4] and in the US the number of such programmes is even bigger.[5] They operate at all levels (national, local, international) and are supported by different stakeholders (*i.e.* public authorities, business, NGOs): they often take the form of awareness campaigns and information tools (guides, brochures), food redistribution programmes, logistical improvements, research and training programmes. As most of these initiatives are very recent, many have not yet been assessed and there is little material about them.

5.2 FEEDING THE HUNGRY

The economic global recession and higher food prices have increased the number of people struggling with food security. As a consequence, recovery programmes and food donations have grown in number. The US Department of Agriculture (USDA) estimates that in 2010 17.4 million US households, *i.e.* 1.2% more than in 2009, encountered difficulty in providing adequate food for all their members.[6]

Food recovery is generally made up of a series of activities where discarded food materials are collected, sorted and converted into other materials or used in the production of new products. Of course, not all food that is lost is suitable for consumption, and assessing the quality of donated or recovered foods is necessary to maintain safety.[7] For example, perishable foods past their 'use by' date, foods in sharply dented or rusty cans, foods in opened or torn containers, and foods prepared, cooked, cooled, or reheated at home (except for baked goods) are not suitable for donations.

Food recovery programmes are run by food banks as well as by governments, retailers, independent farmers, restaurants, non-profit organizations, industries and hotel chains. Some programmes involve a handful of dedicated volunteers in a barely serviceable facility; others are larger, with paid staff and state-of-the-art facilities. Once rescued, food is distributed to agencies and charities that serve hungry people such as soup kitchens, youth or senior centres, shelters and pantries.

Food recovery programmes can offer numerous benefits to businesses and communities. They can:

- Provide wholesome food to needy families in the community.
- Allow recipient programmes to reallocate money spent on food to other services.
- Help communities and businesses meet state and local waste reduction goals.
- Reduce waste going to landfills and decrease the costs and environmental impact of solid waste disposal.
- Allow tax savings for farmers, food manufacturers, retailers, food service operators, and others who donate food.
- Save businesses money otherwise spent on trash collection and disposal fees.

- Create an improved public image for businesses.
- Help sustain local industries and jobs.

For food producers, processors and corporations with food service operations, donating surplus food to the needy can be an excellent way to make use of wholesome excess food. A growing number of businesses have begun to donate their excess food as part of their overall waste reduction strategy. Beyond the environmental and cost savings benefits of donating food, these businesses also have the satisfaction of knowing they have helped feed someone who otherwise might have gone hungry.

Despite a number of positive impacts, food donation schemes have enthusiasts and doubters, especially in the UK.[8] In favour are the redistribution agencies and recipient charities that make use of surplus food. They maintain that food recovery and donations play a vital role in alleviating short-term hunger. Against are some organizations, like Sustain, the British alliance for better food and farming, which affirms that surplus food redistribution should be approached with caution: it can in fact have unwelcome long-term effects like maintaining rather than solving the problems of food poverty. In its report, *Too Much and Too Little*,[9] Sustain argues that private large-scale food redistribution schemes in the US have become institutionalized and a permanent extension of the welfare system and that there are more sustainable ways to address hunger than relying on food hand-outs. Furthermore, retailers are often accused of donating food simply to clean up their image as offenders on social and environmental issues. Tim Lang, Professor of Food Policy at the City University of London and a worldwide expert who has conducted ground-breaking work on food sustainability, often states in his public speeches that food donations cannot really save people from being hungry. What is for sure is that in recent times food donation programmes have been spreading and governments as well as local authorities are trying to enhance donations by providing guidelines and fiscal benefits to the organizations for such initiatives.

In this chapter we focus mainly on food rescue initiatives we have been investigating for the last 2 years. Interesting case studies from four selected countries—US, UK, Denmark and France— are reported and in-depth information on specific ongoing programmes, awareness campaigns, retailer initiatives and activities of charities will be provided. The programmes are different from country to country and distinct in terms of size, organization, management and clientele. As mentioned before, most of these initiatives are very recent: as a consequence most information has been collected thorough websites and online resources.

5.3   FOOD RECOVERY AND FOOD DONATION PROGRAMMES IN THE US

According to the EPA, the US generates more than 34 Mt of food waste each year and food waste represents more than 14% of the total municipal solid waste stream.[10] Of the 34 Mt of food waste generated in 2009, less than 3% was recovered and recycled. The rest—33 Mt—was thrown away into landfills or incinerators.[11] Retailers and restaurants throw away 35 Mt a year, valued at $30 billion. Households are responsible for throwing away approximately $43 billion worth of food (not including plate scrapings, garbage disposal waste, or composting). That comes out to about 14% of what they buy, or 1.28 lb (0.58 kg) of food per household per day.[12] The disposal cost of such food exceeds $1 billion in local tax funds annually.

At the same time, hunger is a serious and complex problem in the US. The economic recession, resulting in dramatically increasing unemployment nationwide, has driven unprecedented, sharp increases in the need for emergency food assistance and enrolment in federal nutrition programmes.

In 1998, about 36 million Americans, including 14 million children, lived in households that suffered from either hunger or food insecurity.[13] A study by Second Harvest, the national food bank network, indicates that an estimated 21 million Americans depend upon charitable food donations to prevent their families from going hungry and that food banks' emergency feeding programmes frequently run out of food before they can serve all the families in need of assistance.[14] Feeding America's last study reported that if merely 5% of food discards were recovered, 14 million additional Americans could be fed each day.[15]

A growing national movement is under way to recover excess wholesome food and distribute it to hungry Americans. To encourage businesses to participate in food recovery, despite the high costs sometimes related to donations, the EPA offers a food recovery challenge for businesses and provides online tools that companies can use to conduct a baseline food waste assessment, create a food recovery plan, and report annual progress.

USDA[16] provides guidance to food rescue organizations on how to use administrative state funds to support their activities. For example, the Redwood Empire Food Bank[17] (REFB) is one of the many organizations that receive funds and distributes food throughout Sonoma County, California. In 2009, REFB received $34 024 in funds from the State of California to support its own food assistance programmes and other 147 community-based charitable organizations.

Some of the most prominent food recovery and donations programmes in the US are described below.

5.3.1 Food Bank Initiatives

The *Second Harvest Food Bank* network[18] was founded in 1965 by a businessman called John Van Hengel who volunteered to feed homeless people at St Mary's Mission in Phoenix, Arizona. Second Harvest is the nation's largest domestic hunger relief charity. The Second Harvest network is comprised of 188 affiliate food bank members providing more than 1 billion pounds (450 million kg) of food and grocery products to 45 000 local charitable agencies. The Second Harvest network provides food to approximately 26 million low-income Americans, including 21 million people at soup kitchens, food pantries, and other emergency feeding sites. Food donations to Second Harvest come from more than 500 national donors, from farmers, local food drives, and the federal government. Second Harvest food banks provide assistance in all 50 states, the District of Columbia, and Puerto Rico, and serve nearly every US county. Each affiliate food bank is local community supported and volunteer-based.

5.3.2 County Programmes

The 'Weekly Pickup—Green Yard Trimmings and Food Scraps Cart' programme, run by StopWaste.org,[19] the Alameda County Waste Management Authority (California), allows households to recycle food scraps, along with food-soiled paper, with other organic yard waste. Currently, food scraps collection is available to approximately 95 000 households. Participation rates are currently being assessed, but the weekly green cart collection and weekly recycling collection makes it easier for Oakland residents to recycle more and waste less.

5.3.3 Retailer Initiatives

Supermarkets are a major organics generator, with over 400 supermarkets in the US generating an estimated 90 600 t of organic material per year. The Massachusetts Department of Environmental Protection (MassDEP) partnered with the Massachusetts Food Association (MFA) has established a voluntary Supermarket Recycling Program Certification (SRPC) programme to promote reducing, recycling, and reusing food waste and other materials called '100 Supermarkets in Massachusetts'.[20] Now, supermarkets in Massachusetts can obtain SRPC status by annually certifying to MassDEP that they have a comprehensive recycling and re-use programme in place. Recycling and re-use programmes can include food donations to local food shelters and diversion of food scraps, cardboard, paper, plants, and wood boxes to composting. In turn, MassDEP provides technical assistance to their supermarket partners in developing their programme, such as the *Supermarket Composting Handbook*. Participating supermarkets save money, and receive positive recognition and regulatory relief.

As of August 2005, 62 supermarkets, 9 hauliers, and 6 composting facilities were recovering organics to achieve a 60–75% recycling rate of food scraps and other organics in Massachusetts. On average, participating supermarkets saved more than $4400 per year per store and collectively recycled 65.9% of supermarkets' total waste stream avoiding $700 000 in reduced disposal costs.

5.3.4 NGO Initiatives

City Slicker Farms[21] is an organization based in West Oakland, California, that runs organic, sustainable, bio-intensive market farms and backyard gardens. The produce from these farms and gardens provides affordable, fresh produce to the local community. City Slicker Farms accepts donated food and yard scraps from West Oakland residents, which is composted and used for their farm and garden needs. In 2005, they diverted close to 20 t of food scraps and yard waste from landfills. City Slicker Farms is currently unable to generate all the compost that they need to run their farming operations through this donation programme, although they are interested in expanding towards a goal of self-sufficiency.

Food scrap recovery programmes like the one run by City Slicker Farms have a number of benefits, beyond simply reducing the amount of food scraps that end up in landfills. They connect an individual household's waste production with food production within the realm of the community.

Households that might traditionally be considered part of 'hard to reach' populations (*i.e.* members of multi-family residences, or those who do not value recycling highly) may be more inclined to participate in food scrap recovery programmes that are built on community relationships. In general, City Slicker Farms' collection programme and other 'non-commercial' food scrap recovery programmes have the potential to complement larger, commercial programmes by reaching out to community members, and by exemplifying how food scraps can be utilized in the sustainable production of fresh produce for the community itself.

5.3.5   Think Tank

The 'Rock and Wrap It Up!' (RWU)[22] programme is a national anti-poverty think tank that arranges the collection and local donation of leftover food and other basic necessities, such as toiletries, from rock concerts, sporting events, hotels, corporate meetings, political rallies and school cafeterias. As a general rule, caterers prepare 10–15% more food than they need for an event, making for a lot of leftovers.

RWU began collecting leftover food in 1991 and was launched nationwide by MTV in 1994. Since its inception, the organization has donated to over 41 000 shelters and places of need, collaborated with 150 bands, 200 schools and universities, and 30 sports franchises, collected more than 100 million lb (45 million kg) of food and fed more than 200 million people.

5.3.6   School Programmes

The Chez Panisse Foundation[23] is a non-profit organization founded by chef and author Alice Waters. In 1995 it launched the

Edible Schoolyard (ESY), a programme consisting of the cultivation of 1 acre (0.4 ha) organic garden and kitchen classroom for urban public school students at Martin Luther King, Jr. Middle School in Berkeley, California. At ESY, students participate in all aspects of growing and harvesting, and prepare nutritious, seasonal meals, making sure no foods go wasted. Students who participate in the ESY programme learn about the connection between their everyday food choices and the health of the community, the environment and themselves. These lessons foster sound nutritional practices, responsible food choices and environmental stewardship.

The approach is to demonstrate the entire 'farm to table' process to students and to procure healthy lunches from local and sustainable sources. To date, a small network of ESY affiliate programmes in cities across the country has been established.

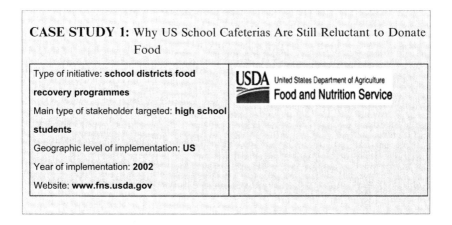

CASE STUDY 1: Why US School Cafeterias Are Still Reluctant to Donate Food

Type of initiative: **school districts food recovery programmes**	USDA United States Department of Agriculture **Food and Nutrition Service**
Main type of stakeholder targeted: **high school students**	
Geographic level of implementation: **US**	
Year of implementation: **2002**	
Website: **www.fns.usda.gov**	

In this section we report on a case study based on research conducted by two Kansas State University researchers in 2002.[24] Lee and Shanklin (2002) had previously reported that only 1.5% of school food service programmes saw the act of donating foods to charity organizations as a way of decreasing food waste.[25] As a way to encourage food recovery in school food service, in 2002 USDA funded 12 school districts to assist them in establishing or expanding food recovery programmes. The population for the study was public school districts located in the USDA Mountain-Plains Region that participated in the National School Lunch

Program (NSLP). The Mountain-Plains Region includes Colorado, Iowa, Kansas, Missouri, Montana, Nebraska, North Dakota, South Dakota, Utah and Wyoming.[26]

One hundred participating school districts were randomly selected from each state, except Utah and Wyoming. The final sample size was 887. A questionnaire was developed to address the availability of food recovery programmes and the food recovery practices used in school food service operations. For school districts that did not participate in food recovery, a series of questions addressing factors that discouraged implementation of food recovery and information needed for initiating food recovery were included.[27]

Despite encouragement from USDA, the participation rate in food recovery was low. Among the 623 responding school districts, only 37 (5.9%) reported that their district participated in a food recovery/donation programme.[28] The 24 school districts with specific food recovery programmes, however, were located in metropolitan or urban areas with populations of more than 2500. The majority (82%) of the school districts without food recovery programmes were located in rural areas in each state. Between one and six schools donated food, except for Missouri, where no school food service programme reported donating food. Among the 37 school food service operations with food recovery donations, 9 school districts donated both perishable and non-perishable, prepared foods and 7 operations donated only non-perishable food items (Table 5.1).

The annual donation ranged from 5 to 5000 lb (2–200 kg) of perishable, prepared food and from 40 to 100 lb (18–45 kg) of non-perishable food items. The frequency of donation may explain the wide range in the quantity of food donated. Fourteen school districts donated food only when schools were closed or when they had special events. Three school districts donated food daily. The school food service programmes donated food most frequently to food banks, community soup kitchens and homeless shelters.

Transportation and labour were most frequently listed as the primary costs incurred by participating districts. For seven school food service programmes no additional costs were involved in food recovery, except the food itself or a telephone call to request an agency to pick up the donation. Additional staff training for food recovery was not a deterrent to participation, since school food service staff already have expertise in safe food handling.

Table 5.1 Food recovery practices of participating schools[a]

Variable	N	%
Frequency of food donation (n = 37)		
Daily	3	8.1
1–4 times/week 4 10.8	4	10.8
1–2 times/month 4 10.8	4	10.8
3–4 times/month 3 8.1	3	8.1
Less than 4 times/year	3	8.1
Only when school closes or special events	14	37.8
Other	6	16.2
Recipient organization[b] (n = 37)		
Food bank	18	48.6
Community soup kitchen	9	24.3
Homeless shelter	7	18.9
Local church	2	5.4
Emergency shelter	2	5.4
Other	9	24.3
Motivation to initiate food recovery (n = 36)		
Reducing edible food discarded	23	63.9
Enhancing public image	5	13.9
Decreasing waste removal costs	3	8.3
Right thing to do	2	5.6
Complying with school policy	1	2.8
Other	2	5.6
Primary cost of programme participation (n = 28)		
Transportation	9	32.1
Labor	7	25.0
Packaging	2	7.1
Volunteer training	2	7.1
No cost	7	25.0
Other	1	3.6
Delivery responsibility[b] (n = 36)		
School food service staff	18	37.5
Staff from recipient organizations	16	33.3
School volunteers	7	14.6
Volunteers from the community	6	12.5
Other	1	2.1

[a]Participating schools were those schools that were involved to some extent in food recovery activities (N = 37)
[b]Respondents were allowed to select more than one choice; therefore, the sum of the percentages exceeds 100
Source: http://docs.schoolnutrition.org/newsroom/jcnm/02spring/lee/

Providing food for the hungry, reducing the amount of edible food discarded and enhancing public image were identified as the most important benefits for participating in food recovery programmes. Challenges faced in implementing food recovery included concern about food safety of donated food and availability of staff and volunteer time. Obtaining administrative support and limited knowledge of food recovery were not identified as challenges by the directors surveyed, who had initiated food recovery programmes (Table 5.2).

Food service directors whose programmes did not participate in food recovery (n = 586) were asked about operational and administrative/regulatory factors that discouraged them from having a food recovery programme. The majority of these directors (64%) indicated that they did not donate food because they did not have enough food to donate. Having little or no food left at the end of service is the goal of all programmes. Approximately 16% of the respondents indicated that they were not aware of available options. Some directors reported that concern about food safety (5.7%), need for special equipment (2.2%) and labour availability (2.1%) were reasons they did not donate food. Lack of information about regulations (34.4%) was the most frequently mentioned

Table 5.2   Benefits and challenges in implementing food recovery of participating schools[a]

Variable	N	%
Benefits (n = 37)		
Providing food for the hungry	25	67.6
Reducing edible food discarded	24	64.9
Enhancing public image of programme	6	16.2
Decreasing waste disposal expenses	4	10.8
Involving students in community service	4	10.8
Increasing awareness about hunger	3	8.1
Other	1	2.7
Challenges (n = 31)		
Food safety	16	51.6
Staff and volunteer time	7	22.6
Employee training	3	9.7
Lack of information	2	6.5
Cost of implementing programme	1	3.2
Other	2	6.5

[a]Participating schools were those schools that were involved to some extent in food recovery activities (N = 37).
[b]The respondents were asked to select two choices; therefore, the sum of the percentages exceeds 100.
Source: http://docs.schoolnutrition.org/newsroom/jcnm/02spring/lee/

reason for not participating. Respondents also were concerned about the cost of programme implementation (18.2%) and legal liability (12.8%).[29]

The respondents rated their knowledge of the Bill Emerson Good Samaritan Food Donation Act as low, regardless of their participation. Only 1 out of the 37 directors with a food recovery programme reported being very knowledgeable about the Act, and 13 of these directors had never heard of the Act. Some directors expressed concern that their staff would overproduce for donation.

The outcome of the project was that, interestingly enough, despite administrative and state support, school food service programmes in the US were, and still are, reluctant to donate food, or may limit their donations to bakery items—the kind of foods organizations need the least.

5.4 FOOD RECOVERY AND FOOD DONATION PROGRAMMES

5.4.1 UK

Proportionately, the UK and Japan have traditionally been among the worst offenders worldwide when it comes to food waste, discarding 30–40% of their food produce annually. In the UK, 6.7 Mt of wasted food amounts to £10.2 billion each year.[30]

The Waste & Resources Action Programme (WRAP) has estimated that the British throw away one-third of the food they purchase, adding up to 4.4 million apples, 1.6 million bananas, 1.3 million pots of yoghurt, 660 000 eggs, 440 000 ready meals, 1.2 million sausages and 2.8 million tomatoes annually. The total of daily waste costs an average home more than £420 a year but for a family with children the annual cost rises to £610.[31]

In their report *The Food We Waste*,[32] WRAP also revealed the extent of Britain's throwaway food culture after sifting through the dustbins of 2138 people who signed up to an audit of food detritus: one in three shopping bags is dumped straight in the bin. Approximately 500 000 t of food waste from UK retailers specifically are disposed of annually, largely to landfill.[33] This is worrying, considering the fact that in Britain food prices rose by 4.7% in 2010 (wheat has increased by up to 11% in the past year), and more pressure is put on domestic budgets.[34] According to the charity

Foodbank, 13 million people across the UK are living on the edge of poverty. In this recession, many people are struggling to make ends meet in London, one of the most expensive cities in the world. Seven million people in the UK are affected by low income, and it is estimated that around 4 million people in the UK cannot afford a healthy diet and do not have enough money to afford fresh fruit and vegetables, or two meals a day.[35]

Britain is committed to reduce the amount of food waste going to landfill by 60% by 2016, as landfill tax is due to rise dramatically.[36] Local authorities spend £1 billion a year disposing of food waste, which leads to the release of methane, a potent climate-change gas. Food waste costs UK consumers £10.2 billion a year.[37] With the added factors of production, transportation and storage it is responsible for 5% of the UK's greenhouse gas emissions.

Wales and Scotland are leading the way in terms of food waste reuse and management. Twenty-two Welsh councils offer a separate food or combined food and garden waste collection. About 60% of Welsh homes have access to food waste services and the Welsh Assembly committed £34 million between 2009 and 2011 to extend collections to every household. Most English councils are lagging behind; only 41% of the 300 English councils responding to the survey collect food separately. An *Independent on Sunday* survey[38] of local authorities has revealed that while the vast majority collect garden waste for composting, less than half pick up food waste separately from general household rubbish.

Among the most prominent food recovery and donations programmes in the UK are the following.

Campaigns Against Food Waste.   The *'Love Food, Hate Waste'*[39] campaign was launched in 2007 by WRAP and claims to have prevented 137 000 t of waste being sent to landfill and saved £300 million. The campaign provides handy tips, advice and recipes for leftovers to help everyone waste less food. On the campaign's website there are tips on how to plan the right portions, which essentials should be kept in the cupboard, fridge and freezer, and many recipes on how to reuse leftovers. In relation to the Love Food, Hate Waste campaign, UK minister Hilary Benn announced plans to end the use of the 'use by' date as it is considered a major cause of food waste. There are plans to replace the 'use by' date by new technologies such as time–temperature indicators, a device or smart label that shows the accumulated time–temperature history of a product and indicates exposure to excessive temperature (and time at that temperature).

'This is Rubbish'[40] is a campaign thought up by a group of Welsh young people, aiming at educating people about the scale of food waste in the UK, and at emphasizing the fact that reducing this waste is the joint responsibility of people and companies. 'This is Rubbish' wants all retailers to be obliged to report on the food waste they generate in their activities, with annual reports audited and published by an independent commission. The campaign has supported events including 'Feeding the 5000' (a mass food waste feast in Trafalgar Square), 'Dining Down to Earth' at the Arcola Theatre (3 course dinner for 100 and a mass fruit salad toss), 'Hungry for Activism' (a food chain game workshop), 'Feast Machynlleth' (community feast on local surplus) and 'Seedy Sunday' (information point at local seed swap).

The *'Dinner Exchange'*[41] is a monthly event organized by a group of vegans in London. It consists of a dinner for 30 people, held in a different place each time. All meals are vegetarian, consisting of three courses and providing a balanced diet influenced by current nutritionist guidelines. Most of the food used for the dinners is food that would otherwise go to waste and is donated by sellers with warehouses at the New Covent Garden market and Earth Natural Foods.[42] All proceeds go to charity and guests are expected to give a donation; the recommended amount is £10. The social aspect of these dinners is important as well, as they become a platform for discussions of environmental and humanitarian issues.

The *'Grow Sheffield's Abundance'* project[43] is a community-led group that harvests surplus fruit from trees around the city and redistributes it locally. In 2010 it embarked on a new venture; borrowing a kitchen and working with volunteers to produce their own brand of chutneys and jams. It is selling bottled goods at local farmers' markets and food events around the city, and any income raised goes back into the project.

Non-Profit Organizations/Food Redistribution Programmes. *Food-Cycle*[44] is a charity, established in 2009, whose aim is to empower local communities and set up groups of volunteers to collect surplus produce locally and prepare nutritious meals in unused professional kitchen spaces (Figures 5.1 and 5.2). The meals are then served to those in need in the community. At present it runs five hubs in London and nine outside London, in Birmingham, Bristol, Cambridge, Durham, Edinburgh, Leeds, Liverpool, Manchester and Norwich.

FoodCycle is also developing community cafes, where hearty meals are served at fair, affordable prices. The cafes are open to everyone around lunchtimes for up to four days each week. All volunteers are trained for accreditation in level 2 food safety in catering and benefit from hands-on training during the cooking sessions.

'Approved Food'[45] is a UK-based food redistribution programme that specializes in selling dry food products—more than 700—that are near or past their 'best before' date (not the 'sell by' date) at a discounted rate through its website. Sales and revenue figures are not available, but the company has received a large amount of mass media publicity, indicating an impact on consumer awareness. The company represents an innovative private-sector approach to avoiding food waste by means of resale.

Figure 5.1   FoodCycle community kitchen. Photos courtesy of FoodCycle.

Figure 5.2   People enjoying food rescued by FoodCycle. Photo courtesy of FoodCycle.

Retailer Initiatives. A number of retailers have already implemented initiatives to reduce food waste as part of the *Courtauld Commitment*.[46] This is a voluntary agreement between WRAP and major UK grocery organizations to minimize packaging and cut food waste. As part of its wider food waste strategy, WRAP, UK governments and industry partners are working together to develop and implement category-specific action plans. Grocery retailers and manufacturers have pledged to work together to cut UK household food waste by 155 000 t or 2.5% total waste before the end of next year.[47]

The *Sainsbury's* supermarket chain[48] has developed a programme to send 42 t of food waste to a biofuel refinery plant in Motherwell, Scotland,[49] where it will be converted into fuel suitable for generating electricity. Each tonne of food waste diverted from landfill by Sainsbury's will generate enough power for 500 homes and will save 3 t of CO_2 compared with fossil fuels. The move marked the first step in a wider UK plan to stop using landfill for food waste by this summer 2011 and stop using landfill for all waste by the end of the year.

Tesco[50] has been often criticized for its policy, bearing an almost 30% share of the grocery retail market. In a bid to reduce waste and improve its image, Tesco is offering a staggered approach to buy one get one free (BOGOF) promotions. Consumers will be able to buy and pick up their second, free product at a later date, instead of having to take both at the same time. BOGOF deals came under fire in the Government's Food 2030 report in August 2010 for encouraging shoppers to buy products they did not need and contributing to high levels of food waste. The Tesco chief pledged that Tesco will become a carbon neutral business by 2050. Other initiatives are currently being designed to reduce the carbon emissions in the retailer's supply chain alongside measures to help consumers reduce their carbon emissions. The Green Clubcard points scheme, which allows each holder to receive points for every bag they do not use, will be extended to encourage carbon-conscious purchasing, and customers can access the Tesco Home Energy and Emissions Service to get advice on home insulation and green energy use.

Furthermore, Tesco's new distribution centre in Widnes, England, will be 100% powered by renewable energy generated from food waste.[51] Through a partnership with logistics company Stobart and food waste recycler PDM Group, the new 500 000 sq ft (46 452 m^2) distribution centre will be supplied with renewable energy from PDM's combined heat and power plant. The UK retailer's project will also reduce CO_2 emissions by about 7000 t annually.

In May 2010 *Morrisons*[52] launched the 'Great Taste, Less Waste' campaign to advise consumers on ways to minimize food waste. The supermarket introduced 'Best Kept' stickers on fresh produce to show how best to store fresh foods for longer at home. This is supported with advice through the greengrocery, butchery and fishmonger's departments. People can also buy one chop, or one sausage if that's all they really need.

The *People's Supermarket*[53] (Figures 5.3–5.5) is a cooperative, owned and run by the people who shop there. Open since June 2010 in central London, its ultimate aim is to bring an end to the big supermarket chains and promote a sustainable way of shopping. Once a membership fee (£25) is paid, members are entitled to a 10% discount on all the produce and sign up for a 4 hour shift working in the shop. The fruit and vegetables, which are laid out on old second-hand tables as they would be in a market or old-fashioned greengrocer's, are sourced from some of the best farmer's markets around. There are also selections of handmade breads (the People's Loaf) and cakes as well as most of the usual foodstuffs. A kitchen within the supermarket allows volunteers to cook (and sell) sandwiches, cakes and pies made with food that would have normally been thrown away by 'traditional' super-markets (Figure 5.6).

Figure 5.3 Street entrance of the People's Supermarket. Photo by Silvia Gaiani.

CASE STUDY 2: FareShare—A UK Charity Relieving Food Poverty

Type of initiative: **national UK charity**

Main type of stakeholder targeted: **people suffering from food poverty**

Geographic level of implementation: **UK**

Year of implementation: **2004**

Website: **www.fareshare.org.uk**

Figure 5.4   Interior view of the People's Supermarket. Photo by Silvia Gaiani.

FareShare is a national UK charity supporting communities to relieve food poverty. Established in 1994 as a project within the homelessness charity Crisis, FareShare aims to help vulnerable groups, whether they are homeless, elderly, children, or other groups in food poverty. It provides quality food—surplus 'fit for purpose' product from the food and drink industry, mainly multinationals like Nestlé—to organizations working with disadvantaged people in the community. Some donors deliver these items to the FareShare warehouses from their depots prior to the point of despatch to their retail outlets as part of their logistics chain. Other donors expect FareShare operatives to visit their retail outlets and collect surplus items on an *ad hoc* basis. The scale of donations also varies, from a lorry-load of a particular item to examples such as a small donation of unsold sandwiches from a single retail outlet.

The main recipients are day services (30%), hostels (27%) and homeless groups.[54] Meat/fish/poultry, followed by fruit, ready meals and ethnically/culturally diverse foods (*e.g.* halal/kosher) are the four most important food categories provided by FareShare. The FareShare franchisees treat 'best before' and 'use by' dates as identical, rejecting food that is past its 'best before' date.

Figure 5.5   Community food in the People's Supermarket. Photos by Silvia Gaiani.

FareShare's main depot is based in London (Figures 5.7–5.10) where 2 staffers and 35 volunteers collect surplus food from 20 supermarkets (including Sainsbury's and Marks & Spencer), check it, store it in refrigerators or on shelves and then deliver it to people in need. In the London area alone 13 000 meals are delivered every week to 62 charities and associations that help people rebuild lives damaged by homelessness, mental illness, drug/alcohol abuse and social exclusion.

Outside London, FareShare is present in 14 locations across the UK and currently runs operations in Aberdeen, Barnsley, Birmingham, Brighton, Bristol, Dundee, Edinburgh, Leeds,

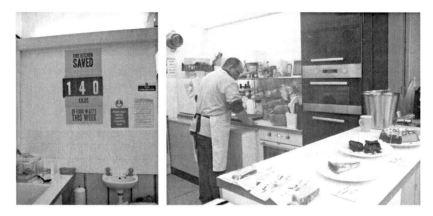

Figure 5.6   Kitchen in the People's Supermarket. Photos by Silvia Gaiani.

Figure 5.7   FareShare—the entrance. Photo by Silvia Gaiani.

Leicester, Llandudno, Liverpool, Manchester and Sunderland. From these locations, 65 towns and cities across the UK are served.

In 2008 the food collected and redistributed by FareShare throughout the UK contributed towards 4.6 million meals.[55] The FareShare Community Food Network has 600 Community Members across the UK receiving food. Every day an average of 29 000 people benefit from the service FareShare provides: 43% of

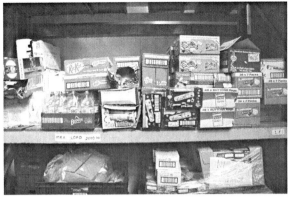

Figure 5.8 FareShare—the depot. Photos by Silvia Gaiani.

Community Members provide breakfast, 46% provide lunch, 36% dinner, and 60% snacks. On average, most members provide each meal 5–6 days per week.

A research project conducted in 2006 by two researchers of the Department of Anthropology, Goldsmiths, University of London reported the items received at Southampton FareShare aggregated over 2 days on the basis of fieldwork (Table 5.3).[56]

Southampton FareShare depends largely on donations from two major retailers, Marks & Spencer, which delivers three times a week from its depot in Thatcham, and Sainsbury's, which offers food for collection from its four local superstores. Depot deliveries are pre-retail food items. The Marks & Spencer items are predominantly pre-packed ready meals and desserts, while the bulk of the Sainsbury's donations is fresh fruit and vegetables.

During the fieldwork period, Sainsbury's offered 185 kg of food donations on one of the fieldwork days and 351 kg on the other (Figure 5.11). Of this 536 kg, 104 kg was rejected by FareShare

Figure 5.9   FareShare depot and van. Photo by Silvia Gaiani.

Figure 5.10   FareShare depot—whiteboards near the entrance. Photo by Silvia Gaiani.

Table 5.3 Sainsbury's food donation to Southampton FareShare (2-day aggregate, February 2006)

Food	Weight (kg)	% of accepted total by weight	% of offered total by weight entering waste stream
Offered	536	—	0
Accepted	432	100	19
Diverted to animal sanctuary	51	12	0
Food and packaging discarded at sorting stage	34	8	6
Distributed to projects	347	80	0

Source: adapted from C. Alexander, C. Smaje, Evaluating Third Sector Reuse Organisations in the UK: Case Studies and Analysis of Furniture Reuse Schemes, Resources, Conservation and Recycling Volume 52, Issue 5, March 2008, pp. 719–730.

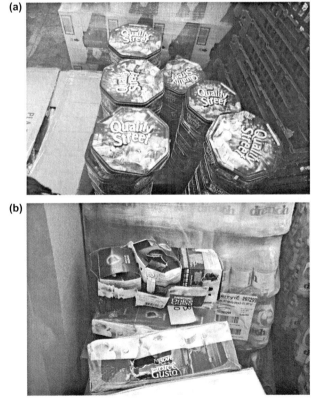

Figure 5.11 FareShare donations: (a) Damaged tins of chocolates. (b) Damaged packets of coffee. Photos by Silvia Gaiani.

operatives at the point of offer on the grounds that it was actually or potentially unfit for human consumption. The remaining 432 kg was accepted and taken to the FareShare depot. Another 85 kg (food that was judged to be unfit for human consumption and packaging removed from food items) was discarded at the depot. Of this, 51 kg was delivered to a local animal sanctuary for animal feed/composting, and the remaining 34 kg were sent to landfill. This left 347 kg to be delivered to recipient projects.

As Table 5.3 shows, 80% of the food accepted by FareShare (and 65% of the food originally offered by Sainsbury's) was ultimately delivered to the recipient projects. The 19% of food offered by Sainsbury's but rejected by FareShare was discarded. In the same fieldwork period, Marks & Spencer delivered 1624 kg of food, or 325 kg per day. This is the thrice-weekly delivery averaged over 5 days, although in this particular week there were only two Marks & Spencer deliveries. Again, this is considerably less than Fare-Share's figures for the 6-monthly average collection weight for Marks & Spencer's of 552 kg per day. On the basis of monthly average figures from the franchise operation, 10 kg of this delivery (packaging and food) was subsequently discarded, the remaining 1614 kg being distributed to projects. These figures are summarized in Table 5.4.

On average, 68% of the food offered by donors to FareShare ends up on the plates of the targeted clients and 58% ends up in the clients' stomachs. Thus, 40% of the food offered returns to the waste stream, the remaining 2% going to the animal sanctuary.

Table 5.4   Marks & Spencer food donation to Southampton FareShare (2-day aggregate, February 2006)

Food	Weight (kg)	% of accepted total by weight	% of offered total by weight entering waste stream
Offered	1642	—	0
Accepted	1624	100	1
Diverted to animal sanctuary	0	0	0
Food and packaging discarded at sorting stage	10	1	1
Distributed to projects	1614	99	0

Source: adapted from C. Alexander, C. Smaje, Evaluating Third Sector Reuse Organisations in the UK: Case Studies and Analysis of Furniture Reuse Schemes, Resources, Conservation and Recycling Volume 52, Issue 5, March 2008, pp. 719–730.

At the supply end, FareShare is a relatively small player within the logistics operations of most retail donors, constituting a disposal option undertaken largely for philanthropic and/or public relations purposes with little impact on the organization's larger environmental responsibilities.

According to the FareShare National Impact Survey conducted in 2008,[57] FareShare has a positive impact on the clients: 92% say that the food they eat at the centre helps them stay fit and healthy. It also has a positive impact on the Community Members, 60% of whom state that FareShare enables them to increase the quality of the food they offer their clients, and 70% agree that FareShare enables them to offer a wider range of food. Furthermore, most of the Community Members are able to save between £1 and £100 per month because of FareShare food, which equates, on average, to approximately 34% of their total food bill. The funds saved are re-invested in various ways like recreational activities, training and life skills. This shows that FareShare is about more than just feeding people; it offers training and education that enhance people's lives. In 2009/10, the food redistributed by FareShare contributed towards more than 6.7 million meals. This redistribution of food helped businesses reduce CO_2 emissions by 12 600 t.

Alongside providing quality food to people in need as part of its work to relieve food poverty, FareShare's goal is to use its experience, skills and position within the food supply chain to provide a programme of education and vocational training opportunities for the vulnerable and disadvantaged people. Through the 'Eat Well, Live Well' programme, it offers education around safe food preparation, diet and nutrition to the staff of the organizations they work with, and to their clients.

FareShare has contributed to and co-organized a series of events like 'Feeding the 5000', an initiative organized by Tristram Stuart, food campaigner and author of *Waste - uncovering the Global Food Scandal*. This event took place on 16 December 2009 in Trafalgar Square, London, where 5000 free meals were served up. The free lunch was made with food that would otherwise have gone to waste, like misshapen carrots and other 'outgraded' fruit and vegetables, which were made into delicious curry and served with surplus bread. Smoothies and cake were also on offer, as well as bags of fruit and vegetables for people to take home and cook themselves. The event aimed to highlight the ease of cutting the unimaginable levels of food waste in the UK and internationally. It provided FareShare with the opportunity to promote the message that 'No Good Food Should Be Wasted'.

FareShare has also supported the 'Big Lunch', when a million people in the UK sat down and had lunch with their neighbours in the middle of their street on 24 July 2010, and the '10 000 Pallet Challenge' in partnership with the Food and Drink Federation, with the aim of getting the industry to get more food that would be sent for waste disposal to people in the community who really need it. Meeting the 10 000 pallet target will enable the charity to double the number of people receiving food to 60 000 individuals a day, increase the number of meals it is contributing towards to 14 million and help businesses reduce CO_2 emissions by 25 000 t.

Since July 2010, FareShare has been part of the London Food Board, a new line-up of food experts who will work to inform the development of programmes to improve Londoners' access to healthy food, boost the economic vibrancy of the food sector and reduce the food system's impact on the environment. It has recently been granted £362 000 from the London Waste and Recycling Board (LWARB) so that over 300 t of surplus food—the equivalent of 800 000 meals—will be diverted from landfill sites each year and distributed to homeless and other vulnerable people.[58]

Future plans for FareShare include:

- The redistribution of 20 000 t of food per year.
- Everyday support for 100 000 vulnerable people through food provision, training and education.
- 6000 volunteer opportunities.
- The provision of nutritious food to 2200 community organizations and charities.

FareShare is also the name of a similar project running in Australia that provides healthy meals for Melbourne's hungry and homeless, using quality food that would otherwise be wasted. In 2008, this project rescued 280 t of food from 80 businesses. More than 1000 volunteers helped give away 560 000 meals for 106 charities.

5.4.2 Denmark

Denmark is one of the most advanced European countries in terms of waste management, prevention and handling. Denmark has chosen to manage household waste and industrial–commercial waste in a comprehensive waste management system, covering both packaging and hazardous waste.

The overall authority in waste matters is the Danish Environmental Protection Agency. Local and regional councils are in charge of the practical administration of waste management.[59] The practical organization of waste management differs from municipality to municipality. Large municipalities typically choose to manage waste themselves, whereas most small municipalities cooperate in inter-municipal waste companies. Further, private companies have been established to operate mainly within collection of household waste and industrial–commercial waste, as well as recycling.

In Denmark there is a general state tax on waste. The waste tax is differentiated so that it is most expensive to landfill waste, cheaper to incinerate it and tax exempt to recycle it: this is why 55% of biowaste is incinerated.[60] Another tax, the so-called 'green' tax, is applied to packaging, plastic bags, disposable tableware and nickel-cadmium batteries.

Despite efficient policies, data regarding food waste in Denmark are alarming. Danish citizens throws away 63 kg (138 lb) of edible food per person every year.[61] An average Danish family with two adults and two children wastes food worth US$1872 (€1341) a year. Danish households account for 89% of the total food waste in Denmark. Interestingly, according to a research conducted by the Danish Agriculture & Food Council in 2009, 48% of Danes are convinced they do not waste food.

Currently there is an ongoing ethical debate on food waste since Denmark, one of the top welfare countries in the world, donates millions of kroner to foreign aid while at the same time wasting tonnes of good edible food. This is aggravated by the fact that nearly 300 000 Danes live below the poverty line.[62]

Among the most prominent food recovery and donation programmes in Denmark are the following:

A la Carte Menu. Hvidovre Hospital in Denmark, led by chef Mogens Pedersen Fonseca, changed how food services are operated to reduce food waste produced by the previously rigid patient catering system.[63] Following on 4 years of extensive work to modify the kitchen and hospital facilities and rethink the cooking strategy, the chef and his 100 employees were able to offer anytime à la carte order options to patients while remaining within budget limitations. The programme has helped the hospital avoid 40 t of food waste per year and the 'à la carte' style encourages portion management; money saved through the initiative has been reinvested to further reduce food waste and improve the quality of hospital food services.

Fish Chips.   In 2009 hospitality and restaurant sector players in Denmark formed a partnership, using state and EU fisheries development funds, to develop an omega-3 rich fish chip product from otherwise inedible fish waste. At the end of 2009, the team was in the final stages of testing, having already negotiated agreements with manufacturers and buyers. While concrete results are not yet available, given that over 50% of fish is discarded as inedible waste in Denmark this seems to be an excellent use for a product that would otherwise be food waste.

CASE STUDY 3: 'Stop Wasting Food' Movement, a Danish Consumer Movement

Type of initiative: **consumer movement**

Main type of stakeholder targeted: **all Danish population**

Geographic level of implementation: **Denmark**

Year of implementation: **July 2008**

Website: **www.stopspildafmad.dk**

Stop Wasting Food is a consumer movement founded to raise public awareness about the food waste subject and to reduce food waste in Denmark. The Stop Wasting Food movement aims at inspiring Danish citizens to take action and donate their good and fresh leftovers to shelters for homeless people.

The movement was established by Selina Juul in 2008 (Figure 5.12). Selina, a graphic designer by profession, decided to take action against food waste after having volunteered for Greenpeace and having read article after article about the food waste issue in Denmark. She now dedicates most of her time to the cause of fighting food waste.

Figure 5.12 Stop Wasting Food in the news (Selina Juul, the founder, is on the left). Photo courtesy of Stop Wasting Food and Rema 1000.

The Stop Wasting Food movement is a privately based and 100% apolitical, non-religious and non-profit organization. Membership is free and at the moment (February 2011) there are more than 7000 members including members of the Danish Parliament, EU politicians, top Michelin chefs and major Danish food personalities. The targets are the entire Danish population, specifically the younger part of the population as older people (60 +) are more mindful about not wasting food.

The movement operates through campaigns in schools, public lectures and seminars (Stop Food Wasting was at COP 15, the United Nations Climate Change Conference, in December 2009 and is frequently invited by the Danish parliament to make presentations and hold public speeches), but mainly raises awareness through the use of information and communication technologies

(Figure 5.13). News about the movement has been reported in over 60 media outlets including radio, TV and newspapers. The movement has a Youtube channel, a Facebook profile, a website and a blog. If people Google 'food waste' (madspild) in the Danish version of Google, the Stop Wasting Food movement is at the moment the leading search result with over 101 000 hits.

In collaboration with Wise Banyan Tree, Stop Wasting Food has launched the Danish version of the iPhone application 'Green Egg Shopper' which helps organize shopping.[64] Green Egg Shopper helps save money and reduce food waste in three simple steps:

Figure 5.13   Stop Wasting Food—awareness campaign on food waste in Copenhagen. Photo courtesy of Stop Wasting Food.

- By creating shopping lists, thus helping to plan in advance and avoid impulse buying.
- By setting use by/best before/expiry dates of perishables as the customer picks items while shopping.
- By checking the list of items near expiry any time.

Stop Wasting Food has a constructive approach to the problem of food waste: its awareness campaign doesn't point fingers or throw pictures of hungry children in people's faces, condemning their food shopping habits. It suggests simple actions that people should put into practice in their daily life (*e.g.* 'when shopping, purchase just the right quantity needed, keep the leftovers and use them for the next day, or donate them to a public kitchen').

It has also recently launched a doggy bag campaign: all restaurants in Denmark can download a doggy bag campaign pack from Stop Wasting Food's website. It has also been successful in introducing, in cooperation with retailers like REMA 1000, a reduction in food packaging—which also causes food waste, as up to 25% of contents are wasted—and the availability of smaller portions of vegetables and meat in supermarkets (60% of Danish single people would like to have smaller portions in shops). The Stop Wasting Food movement also runs national Go-Card campaigns (Go-Cards are free postcards that are available in every café, restaurant and cinema in Denmark).

The movement has recently started a cooperation with the Danish Government and was invited by the former Danish Minister for the Environment Troels Lund Poulsen to participate in the Advisory Board for Waste Prevention Campaign 2010.

Selina Juul is currently working on the production of a leftovers cookbook in cooperation with selected Danish top chefs which will be published by Gyldendal Publishing, Denmark's biggest publishing house. The book will contain recipes for delicious dishes made with good leftovers and much more.

5.4.3 France

Although there are still no precise data about food waste in France, according to a study conducted by the French Environmental Agency (ADEME) 20 kg of untouched, still wrapped food is thrown away per person per year.[65] At the same time 3.3 million people in France in 2009 suffered from food insecurity, including elderly people, unemployed people, homeless, disabled, single mothers, and also, more and more often, young people: of these 3.3 million people, 1.3 million are workers with an average salary of less than €600 per month.

According to figures from the French Federation of Food Banks, in 2009 more than 92 000 t of food (of which 46 400 t would otherwise have been discarded) were used by food recovery programmes to feed about 740 000 people. Among the most prominent food recovery and donations programmes in France are the following:

Awareness Campaigns. ADEME, the French environmental agency, has launched an awareness campaign aimed at informing households about waste production and prevention. This has been in place since 2005 and uses multiple communication channels (online resources, a TV spot, radio broadcasts). The website offers practical tips for food waste related reduction at home and while shopping. No specific results are available but the stated goal is to target the reduction of waste produced annually in France via individual adoption of behavioural changes.

'*Manger autrement dans les collèges*' is a programmes set up by the General Council of Bouches du Rhone and the association 'De mon assiette à notre planète'. It has been in place since 2006 and has organized workshops with the aim of sensitizing students to the quantities of food thrown away in canteens and cafeterias in schools. Eighty-one per cent of college students who participated in these workshops now state that they pay more attention to reducing waste.

CASE STUDY 4: Association Nationale de Dévelopement des Epiceries Solidaires (A.N.D.E.S)

Type of initiative: **solidarity grocery store**

Main type of stakeholder targeted: **people with low purchasing power**

Geographic level of implementation: **France**

Year of implementation: **April 2000**

Website: **www.epiceries-solidaires.org**

A.N.D.E.S, the national association of solidarity grocery stores, is a food aid organization created in France in 2000. Founded by Guillaume Bapst, A.N.D.E.S and its network of 'solidarity grocers' are addressing the problem of hunger and malnutrition in France, as part of a broader objective to re-socialize people living below the poverty line, give them choices as consumers, and help them better manage their daily living.

Through efficient central management, tight supply chains, and economies of scale, the grocery stores (Figure 5.14) provide standard-quality products to low-income families while remaining largely sustainable. Most important, local stores stocked with a variety of groceries offer families choices (being deprived of choice is a form of exclusion) and enjoyment in purchasing their food, giving consumers a sense of autonomy while encouraging a balanced diet.

By giving people a choice of the products they like and by making them pay for it like any other client, solidarity grocery stores aim to break dependency on charity and to improve people's eating habits. Central to the A.N.D.E.S idea are affordability (prices are 10–30% lower than market prices) and liberty to choose between products. This allows those in need to maintain

Figure 5.14   An A.N.D.E.S grocery store. Photo courtesy of A.N.D.E.S.

their self-respect and in some cases, helps them to reverse a dependency on state or charitable assistance.

The idea of creating these paying solidarity grocery shops grew out of Guillaume Bapst's reflection on the nature of 'assistance' and in particular, on the observation that 'a donation which cannot be reciprocated belittles the recipient especially when there is no possibility to give something in return'.[66] The money saved using this model has to be re-invested into a project designed by both the social worker and the client. These projects include children's scholarships, car repairs to commute to work, or debt reduction. This model also helps people to learn how to manage their monthly income better.

In addition to being a better source of food, solidarity grocery stores have become a venue for a variety of other services for rehabilitation and integration. Special efforts are made to give clients a renewed interest in preparing food and in eating a healthy and balanced diet: nourishment becomes pleasurable and provides the occasion for social interaction. In addition, other activities are

Figure 5.15 A cooking activity with children in an A.N.D.E.S grocery store. The program is called 'La Compagnie des Gourmands': the idea is to gather parents and their children and to prepare a meal together. The aim is to promote a healthy diet, have fun and reinforce the family link. Photo courtesy of A.N.D.E.S.

organized, from cooking classes to workshops for parents and children that help reinforce family ties (Figure 5.15).

While each shop is unique, they are all managed centrally and professionally and are built around the core value of respect for the autonomy of the clients they serve. Since competitive pricing is one of the main problems for these stores, there is a centralized logistics and purchasing centre that helps to ensure a constant supply of goods at low prices without sacrificing quality. By buying in bulk, A.N.D.E.S takes advantage of the best prices and lowest tariffs and contributes to surplus public funds at the local level. This logistics platform is about to be set up in Angers.

Strategic partnerships with large brands, such as Nestlé, Unilever, Ferrero and Poulets de Loué have also been established. These partnerships allow A.N.D.E.S to benefit from the best prices and in some cases, receive goods for free (under the French tax framework it is possible for companies to deduct 60% of the value of the products from their taxes). A.N.D.E.S is working to involve large national distributors as partners in their own distribution chain. Because they have a wide presence and a huge negotiating power, they would be strong allies in helping to scale up the A.N.D.E.S programme. By positioning A.N.D.E.S as a partner that will bring thousands of customers into their stores, rather than a competitor, Bapst was recently able to sign an agreement with Carrefour—a large step forward for A.N.D.E.S' national development.

Access to the A.N.D.E.S grocery stores is not permanent, but depends on the plan the client has designed with social workers and the grocery store manager. In most cases, the arrangement involves reducing client debt. This is essential, because it helps clients learn how to better manage their weekly or monthly budget, and teaches them how to make the right choice between two products or begin saving for future expenses, such as education. In the A.N.D.E.S grocery stores, three out of four clients achieve their intended goal.

Although the grocery stores are primarily to sell food, they provide more than that. Each store becomes a creative tool for developing social links to help buyers address other issues, such as health, housing, work and education. In this way, the solidarity groceries are an avenue to social inclusion and to the restoration of economic citizenship. Guillaume Bapst believes it is essential for his customers to contribute to each store by participating in its management and helping to shape and lead the workshops held there.

Some of the most successful client-led workshops have focused on cooking, teaching people what and how to cook for a healthy and balanced diet. From time to time professional cooks are featured, but more often participants are eager to share their recipes and methods.

A.N.D.E.S is also carrying out a fight against food waste and has recently presented a report, in collaboration with France Nature Environment, on how to reduce fruit and vegetable waste, by collecting unsold products on wholesale markets (Figure 5.16).

Today, the A.N.D.E.S network contains approximately 60 grocery stores and serves hundreds of households, almost 24 000 people per year. With a budget of €160 000 for 2006, the association is developing tools (software, training sessions and sharing of

Figure 5.16 Fighting against fruit and vegetable waste and promoting social inclusion in the A.N.D.E.S workshop in Perpignan (France). Photo courtesy of A.N.D.E.S.

Figure 5.17 Every year A.N.D.E.S collects 10 000 litres of milk at the Paris International Agricultural Show. The milk is then delivered to food aid organizations in and around Paris. Photo courtesy of A.N.D.E.S.

best practice) to help manage stocks and is building strategic partnerships with the food industry (Figure 5.17). The aim is to develop more and more groceries of this kind in France, beginning with the most economically devastated areas, like the more than 700 Zones Urbaines Sensibles. There are also plans to enter remote areas in rural zones where more and more poor people live. Soon A.N.D.E.S will broaden the range of products provided, such as furniture and appliances.

5.5 COMPREHENSIVE INTERNATIONAL APPROACHES TO FOOD WASTE

Most of the initiatives presented above have been personally investigated and analysed by the authors. During the last two years (2008–2010) we have carried out intensive researches on food waste and on initiatives aiming at reducing it.

What we basically found out is that, although some countries are doing better than others in terms of food waste reduction and management, the general approach to food waste is changing and thousands of interesting projects all over the world are contributing to raising awareness and promoting actions to lower consumption and increase resource efficiency.

A new model of prosperity for an environmentally degraded and poverty-stricken planet is currently looked for. Efforts to develop systems that circulate materials through recovery and re-use are currently being incentivized, and multisectoral partnerships—like the one established by the Interreg Project GreenCook,[67] financed by the European Commission and involving 12 partners among retailers, production companies, multinationals, schools, municipalities, experts from northern Europe—are encouraged and promoted.

Increasingly, consumers understand that they are paying high prices because of previous unsustainable consumption behaviour and disposal practices that have led directly to environmental clean-up costs and health maintenance costs.[68] Calculations suggest that in order to achieve the twin objectives of environmental protection and social equity, the developed world may need to cut its use of materials by about 90%.[69]

The (food) waste management agenda has recently begun to make a transition, and the focus is shifting from waste towards the sustainable use of natural resources and consistent reduction in consumption.[70] This is clearly evident in the increasing number of resource management strategies that are now replacing waste management strategies. These strategies not only consider physical materials but also seek to reduce the energy required to treat and manage waste and generate an overall reduction in carbon emission. This new focus is encouraging engagement along the entire process chain for products—design, production, and consumption—before finally addressing waste management.

Some very powerful theories are emerging: in particular, the cradle to cradle approach, the zero waste strategy and the degrowth theory.

5.5.1   Cradle to Cradle Approach

Promoted and invented by the German chemist Michael Braungart and the American designer-architect William McDonough,[71] the 'cradle to cradle' approach aims to design every product in such a way that at the end of its lifecycle the component materials become a new resource. Braumgart and McDonough start from the assumption that humans are the only species that produces landfills and that natural resources are being depleted on a rapid scale while production and consumption are rising. Consequently they want to change the way we produce and build. They are calling for the transformation of human industry through ecologically intelligent design. If waste were to become food for the biosphere or the technosphere (all the technical products we make), production and consumption could become beneficial for the planet. The cradle to cradle approach is a comprehensive framework, obviously focusing not only on food, but on all aspects of human production.

5.5.2   Zero Waste Strategy

The basic assumption of the cradle to cradle approach is the same which is at the base of the zero waste strategy, already briefly mentioned in this book. Zero waste is a goal, a process, a way of thinking that profoundly changes the human approach to resources and production. Zero waste is not about recycling and diversion from landfills: it is about restructuring production and distribution systems to prevent waste from being manufactured in the first place. What materials are still required in these redesigned, resource-efficient systems will be recycled right back into production. Zero waste requires preventing rather than managing waste. It emulates natural systems where everything that wears out or dies becomes food or shelter, however temporarily, for something else, giving rise to a vibrant yet efficient flow of energy and resources. Zero waste visualizes the economy as a circular or spiral system in which every part supports and affects every other. It seeks to replace the current outdated linear economic and production system, which does not recognize the interconnectedness of impacts and the trail of wastes left behind.

5.5.3   Degrowth Theory

On the other hand, the degrowth theory, whose origins can be traced back to *The Entropy Law and the Economic Process* written in 1971 by Nicholas Georgescu Roegen,[72] and is now fiercely supported by Serge Latouche and others, suggests changes not only in development patterns but also in individual actions. Society as a whole should consume less but better, produce less waste and recycle more.

The message these theories try to convey is that we should stop thinking of waste as 'waste' and see it instead as a valuable resource for society. In this sense waste ceases to exist.

In the next chapter we describe an Italian initiative, the Last Minute Market project, which is a 360° action against waste and considers waste as a resource. It was originated by one of the authors, Andrea Segrè, in 1998 and since its establishment it has achieved impressive outcomes in food waste reduction, redistribution and management.

REFERENCES

1. M. Betts and M. Burnett, *Study on the Economic Benefits of Waste Minimisation in the Food Sector*, Evolve EB Ltd, 2007.
2. J. Claridge, *Gleaning and Food Recovery as Tools to reconnect at the Local Level*, Footprint Choices, 2010, available at: http://www.slowmovement.com/gleaning.php
3. http://www.epa.gov/osw/conserve/materials/organics/food/tools
4. European Commission, DG Environment, *Preparatory Study on Food Waste across EU27*, Final Report, October 2010.
5. According to USDA it is estimated that there are 150 000 private programs helping to feed the hungry http://www.foodprotect.org/media/guide/Food-Recovery-Final2007.pdf
6. U.S. Department of Agriculture, *Waste Not, Want Not: Feeding the Hungry and Reducing Solid Waste through Food Recovery* (Publication No. 530-R-99–040), 2010.
7. K.S. Kantor, K. Lipton, A. Manchester and V. Oliveira, *Estimating and Addressing America's Food Losses.* Economic Research Services, United States Department of Agriculture, 1997, available at: http://www1.calrecycle.ca.gov/ReduceWaste/Food/FoodLosses.pdf
8. http://www.guardian.co.uk/society/2003/jul/09/homelessness.guardiansocietysupplement
9. http://www.sustainweb.org/about/consultation_document_on_the_creation_of_sustain/
10. Environment Protection Agency, *Industrial and Commercial Waste Survey*, 2002/3, available at: http://www.environment-agency.gov.uk/subjects/waste/1031954/315439/923299/1071046/?version=1&lang=e; 2004
11. http://www.epa.gov/epawaste/conserve/materials/organics/pubs/wast_not.pdf
12. J. Morris, A. Donkin, P. Wonderling and E. Dowler, *A Minimum Income for Healthy Living*, J Epidemiol Community Health 2000;54(12), pp. 885–9.
13. *Report on Household Food Insecurity in the United States*, 1995–1998, USDA Food and Nutrition Service and Economic Research Service, 1999.

14. Second Harvest Report, *Hunger 2009: The Faces & Facts.*
15. U.S. Department of Agriculture Food and Nutrition Service, *The Emergency Food Assistance Program* (TEFAP), 2010.
16. K. Chhabra, *Running Head: Food Waste Reduction by the Implementation of Food Recovery Programs*, San Jose State University, 2010.
17. Redwood Empire Food Bank, *Financial—Annual Reports 2008–2009*, available at www.refb.org
18. http://www.thefoodbank.org
19. http://www.stopwaste.org
20. http://www.mass.gov
21. http://www.oaklandfoodsystem.pbworks.com
22. http://www.rockandwrapitup.org
23. http://www.chezpanissefoundation.org
24. http://docs.schoolnutrition.org/newsroom/jcnm/02spring/lee/
25. K. Lee and C. W. Shanklin, *Food Recovery: A Win-Win Solution for School Foodservice and the Community*, Journal of Child Nutrition and Management, Issue 1, Spring 2002.
26. U.S. Department of Agriculture. (2002). USDA announces food recovery funding for 13 school districts (Release No. 0282.98), available at http://www.fns.usda.gov/fns/MENU/GLEANING/SUPPORT/0282.txt
27. American School Food Service Association, *ASFSA announces Community Kitchen Pilot Schools*, May 21, 2002. available at http://www.asfsa.org/newsroom/pressreleases/ckpilots.asp
28. P.L. Fitzgerald, *When You Waste Not, They Want Not*, School Foodservice & Nutrition, 54(10), 2000, pp. 28–30, 32, 34.
29. U.S. Department of Agriculture, *Best Practices for Food Recovery and Gleaning in the National School Lunch Program* available at: http://www.fns.usda.gov/fdd/MENU/ADMINISTRATION/IMPROVEMENTS/gleaning/gleanman.pdf
30. Department of Environment, Food and Rural Affairs (DEFRA), *Waste strategy for England*, 2007.
31. WRAP, *The Food We Waste: Food Waste Report* (study commissioned from Exodus Market Research and Wastes Work), 2008, available at: http://wrap.s3.amazonaws.com/the-food-we-waste.pdF
32. http://www.wrap.org.uk/retail_supply_chain/research_tools/research/report_household.html

33. C. Alexander and C. Smaje, *Surplus Retail Food Redistribution: An Analysis of a Third Sector Model in Resources*, Conservation and Recycling, Volume 52, Issue 11, September 2008, pp. 1290–1298.
34. http://www.thenewamerican.com/index.php/economy/commentary-mainmenu-43/5512-similar-inflation-patterns-in-us-and-britain
35. *Report on Monitoring Poverty and Social Exclusion*, The Joseph Rowntree Foundation, 2007.
36. http://www.guardian.co.uk/society/2003/jul/09/homelessness.guardiansocietysupplement
37. S. Thankappan, *From Fridge Mountains to Food Mountains. Tackling the UK Food Waste Problem* in Ethics, Law and Society, Vol.3, ed. J. Gunning and S. Holm, Ashgate Publications, 2009.
38. http://www.independent.co.uk/environment/green-living/we-bin-10-wembleys-full-of-food-a-year-ndash-what-a-waste-of-energy-2173989.html
39. http://www.lovefoodhatewaste.com
40. http://www.thisisrubbish.org.uk
41. http://www.thedinnerexchange.zzl.org
42. http://www.earthnaturalfoods.co.uk
43. http://www.growsheffield.com
44. http://www.foodcycle.org.uk
45. http://www.approvedfood.co.uk
46. *Waste and Resources Action Programme*, The Courtauld Commitment, available at: http://www.wrap.org.uk/retail/courtauld commitment/index.html; 2005.
47. http://www.foodnavigator.com/Financial-Industry/Grocers-pledge-to-cut-UK-food-waste
48. http://www.sainsburys.co.uk
49. http://www.businessgreen.com/bg/news/1803055/sainsburys-food-waste-biofuel
50. http://www.tesco.com
51. http://www.environmentalleader.com/2010/02/25/tesco-powers-dc-with-food-waste
52. http://www.morrisons.co.uk
53. http://www.thepeoplessupermarket.org
54. T. Lowe, *The FareShare Approach to Food Waste Reduction*, available at: http://www.resourcesnotwaste.org/members/conf-application-form/Conf-presentations/ 15.TonyLowe.pdf; 2007.

55. FareShare, *Community Food Network. National Impact Survey Summary*, available at: http://www.fareshare.org.uk/pdf/impact survey 05.pdf; 2005.
56. C. Alexander and C. Smaje, *Evaluating Third Sector reuse Organisations in the UK: Case Studies and Analysis of Furniture Reuse Schemes* in Resources, Conservation and Recycling Volume 52, Issue 5, March 2008, Pages 719–730.
57. http://www.fareshare.org.uk/pdf/NIS2008keyfindings.pdf
58. http://www.london.gov.uk/media/press_releases_mayoral/%C2%A3362000-boost-charity-stop-surplus-food-being-dumped-landfill
59. Danish Environmental Protection Agency, *Report on Waste in Denmark*, 2010.
60. http://www.europarl.europa.eu/news/public/story_page/064–78553–190–07–28–911–20100709STO78533–2010–09–07–2010/default_en.htm
61. Danish Agriculture & Food Council, Danish Ministry of the Environment, Eurostat—2006, 2008, 2010.
62. http://www.green-alliance.org.uk/uploadedFiles/Publications/CPPWDenmark.pdf and Danish Labour Movement—2010.
63. http://www.hvidovrehospital.dk/NR/rdonlyres/8A9A2230-EC8E-4B93-ACA2-BC3B4DD6DBC8/0/Final_paper_af_Toubro.pdf
64. http://www.prweb.com/releases/2010/12/prweb4916614.htm
65. ADEME, Dossier de Presse, *Le Gaspillage Alimentaire au Coeur de la Campagne Nationale Grand Public sur la Réduction des Déchets*, 15 November 2010.
66. M. Mauss, *Essai sur le Don. Forme et Raison de l'Échange Dans Les Sociétés Erchaïques*, 1ᵃ ed. 1925.
67. www.green-cook.org
68. M. Fehr, M. Calcado and C. Romão, *The Basis of a Policy for Minimizing and Recycling Food waste*. Environ Sci Policy 2002;5(3), pp. 247–53.
69. E. Weizsacher, *Doubling Wealth, Halving Resource Use—the New Report to the Club of Rome*. Earthscan, 1998.
70. J. Von Braun, *The World Food Situation: Driving Forces and Required Actions*, International Food Policy Research Institute (IFPRI), Washington D.C, Food Policy Papers. available at: http://www.ifpri.org/pubs/fpr/pr18.pdf; 2007 [last accessed Sep 25, 2010].
71. http://www.mcdonough.com/writings/c2c_design.htm
72. N. Georgescu Roegen, *The Entropy Law and the Economic Process*, 1971.

CHAPTER 6

Last Minute Market—a 360° Action Against (Food) Waste

In short, if your basket is empty and the others are full, you can stand this up to a point: then an overpowering envy begins to eat at you, and you can't resist. So Marcovaldo, after having instructed his wife and children not to touch anything, made a sharp turn down a side aisle, out of the sight of his family, where he grabbed a package of dates off a shelf and dropped it into the basket. All he wanted was the pleasure of being able to walk around with it for ten minutes so that he, too, could show off his purchases the way the others did, and then return the package to where he had taken it. This package, together with a red bottle of hot sauce, plus a bag of coffee and a blue pack of spaghetti. Marcovaldo was convinced that, if prudent, he could enjoy for at least fifteen minutes the happiness of someone who can choose a product without having to pay even a cent . . .

Marcovaldo al supermarket
[Marcovaldo at the Supermarket]
(*Italo Calvino, 1963*)

6.1 HOW FOOD HABITS HAVE CHANGED IN ITALY

Eating habits have deeply changed over the past few decades in Italy. Greater prosperity and the resulting greater availability of

Transforming Food Waste into a Resource
By Andrea Segrè and Silvia Gaiani
© Andrea Segrè and Silvia Gaiani, 2012
Published by the Royal Society of Chemistry, www.rsc.org

food have been instrumental in modifying Italian eating habits, mainly by skewing the national diet towards excessive consumption.[1] Previously, the tendency, particularly in rural areas, was to use a much greater range of unsophisticated foods. Vegetables were the main sources of proteins as cheese, eggs and meat were less consumed. The staples were always accompanied by vegetables and fruits, supplementing the basic starch products with proteins and fibres and providing sources of vitamins.[2]

The postwar years in Italy saw a rapid transition from a modest diet to a boom in sophisticated food. As a result of the economic upturn, more people started to eat meat every day and to have biscuits and pre-wrapped pastries or buns for breakfast. Within a very short period of time, Italians made the transition from the vegetable soups and home-made cakes to pre-cooked meals and fast foods.[3]

As a result of the modern way of living, demand for processed, added-value foods increased, driven by young people and families with children who had less time for cooking from scratch and a growing inclination to spend their spare time on leisure and other pursuits.[4]

A remarkable change has taken between 1980 and 2010. The share of household expenditure allocated to food has greatly diminished while the total spending on food service has increased. According to ISTAT data on Italian household consumption, the average monthly expenditure for consumption outside the home (in restaurants, canteens, bars), rose from €38.37 in 1985 to €72.82 in 2005.[5] In 2010 the average household expenditure in food services was around €90 per month in the north of Italy, but under €50 in the south.[6]

If on one hand the quantity and quality of the food has improved, on the other there has been a marked increase in nutritional disorders caused by overeating. People tend to eat more than they need to, and as a consequence 67% of the Italian male population, 55% of the female population and one-third of children between the age of 6 and 11 are overweight.[7] Every Italian has about 3700 kcal per day at his/her disposal: according to FAO estimates the proper amount should be about 1800–2200 kcal.[8]

In 2009, according to a survey of household consumption,[9] 2.6 million households (11.1% of total households) or 7.6 million people (about 13.13% of the total population) were considered as poor.

In October 2009, the weekly shopping expense per capita was estimated at €44.58, 0.54% more than in 2009 (Tables 6.1 and 6.2). This figure was calculated by researchers of the Faculty of Agriculture of the University of Bologna, who developed a monitoring price index called Carocibo.[10] Although apparently this was not a significant increase, it should be noted that between 2008 and 2009 the ISTAT index of consumer prices increased by only 0.30%. The rise in food prices was thus higher, albeit only slightly, and there was an increase in the gap between the price levels of production and consumption.

The figures show that the income of families living in the southern regions in 2008 and 2009 was much lower than those

Table 6.1   Cost of an average weekly diet (CAROCIBO, October 2009)

| Food | Weekly quantity (g) | Food cost (cents) | | Variation % (Oct 09/ Oct 08) | Impact of single food expenditure on weekly shopping | |
		Oct. 2008	Oct. 2009		Oct. 2008	Oct. 2009
Extra virgin olive oil	280.0	157.36	151.76	−3.56	3.55	3.40
Pasta	800.0	132.00	129.60	−1.82	2.98	2.91
Fresh milk	1050.0	155.40	153.30	−1.35	3.50	3.44
Sugar	70.0	6.79	6.72	−1.03	0.15	0.15
Cow's milk cheese	200.0	191.60	190.20	−0.73	4.32	4.27
Mineral water	10500.0	430.50	430.50	0.00	9.71	9.66
Apples	4000.0	692.00	692.00	0.00	15.61	15.52
Parmesan cheese	70.0	111.72	111.86	0.13	2.52	2.51
Bread	700.0	183.40	184.10	0.38	4.14	4.13
Chicken	400.0	178.40	180.40	1.12	4.02	4.05
Fresh meat	200.0	284.00	287.40	1.20	6.41	6.45
Coffee	110.0	107.14	108.46	1.23	2.42	2.43
Biscuits	385.0	135.52	137.45	1.42	3.06	3.08
Wine	1750.0	327.25	332.50	1.60	7.38	7.46
Ham	100.0	217.90	221.80	1.79	4.91	4.98
Eggs	200.0	28.60	29.20	2.10	0.65	0.66
Salad	875.0	147.88	152.25	2.96	3.34	3.42
Tomatoes	700.0	117.60	122.50	4.17	2.65	2.75
Rice	320.0	72.96	78.40	7.46	1.65	1.76
Total (€)		44.34	44.58	0.54	100.00	100.00

Source: Last Minute Market—Econometrica.

residing in the north and consequently also the average expenditure on food was lower in the southern part of the country (Table 6.2). The average diet was more expensive in regions like Valle D'Aosta, Liguria and Veneto and cheaper in regions like Umbria, Puglia and Campania.

Over the past 5 years the prices that have increased the most are those of bread, meat, fish, vegetables and fruits. Currently the poorest families spend up to 42% of their income on food, nearly double the national average (19%, equal to €467 per month).[11]

It is worth noting that the proportion of expenditure for mineral water is 10%. In October 2009, a person with a dietary intake of 2300 kcal a day and drinking 1.5 L of water per day, spent an average of €4.3 per week on mineral water, the equivalent of 9.66% of his weekly food expenditure (the average price of 1.5 L of tap water would be €0.0015).

Table 6.2　Cost of an average weekly diet in each region (CAROCIBO, October 2009)

Regions	City	Weekly cost (€)		Var % (Oct. 09/ Oct. 08)
		Oct. 2008	Oct. 2009	
Abruzzo	L'Aquila	NA	NA	—
Umbria	Perugia	37.77	37.96	0.50
Puglia	Bari	38.00	38.23	0.61
Trentino Alto Adige	Trento	40.28	40.53	0.62
Campania	Napoli	40.52	40.72	0.49
Tuscany	Firenze	41.22	41.44	0.53
Molise	Campobasso	41.65	41.90	0.60
Basilicata	Potenza	42.90	43.11	0.49
Calabria	Reggio Calabria	43.79	44.04	0.57
Piedmont	Torino	44.83	45.09	0.58
Sicily	Palermo	45.41	45.68	0.59
Lazio	Roma	45.68	45.96	0.61
Sardinia	Cagliari	45.79	46.03	0.52
Lombardia	Milano	46.29	46.56	0.58
Marche	Ancona	46.43	46.71	0.60
Friuli Venezia Giulia	Trieste	46.71	46.99	0.60
Emilia Romagna	Bologna	47.19	47.39	0.42
Veneto	Venezia	47.77	48.07	0.63
Liguria	Genova	48.54	48.84	0.62
Valle D'Aosta	Aosta	51.57	51.87	0.58

NA, not available.
Source: Last Minute Market—Econometrica.

6.2   FOOD WASTE IN ITALY

Since 1974, food waste has doubled in Italy. Currently Italy loses the same amount of food along the supply chain that countries such as Spain consume in a year.

Researchers at the Faculty of Agriculture of the University of Bologna, in Italy, have estimated that along the whole supply chain nearly 20 290 767 t of food are thrown away every year, with a value of €3.7 billion: food waste constitutes mainly milk, eggs, cheese and yogurt (39%), bread and pasta (15%), meat (18%), and fruits and vegetables (12%).[12] These figures do not include the 'black hole' represented by food waste originating in canteens, hospitals and schools and in all-inclusive hotels or resort buffets.

These data were obtained by comparing the amount of food that every Italian has at disposal by product type—as reported in the FAO food balance sheets—with the food consumption per capita per day, provided by INRAN.[13]

In Table 6.3, data showing the difference between the amount of food available and that consumed are reported. The percentage of food wasted ranges between 26% and 48% and differs from product to product. Vegetables, fruits, alcoholic drinks and meat are subject to a higher percentage of waste than cereals (36%) and fish (26%) which seem to have a more efficient chain.

The annual amount of food waste along the whole food supply chain has been estimated as follows:

- **Waste at field level**: in 2009, 3.25% (17 700 586 t)[14] of agricultural production, corresponding to a value of €4 billion, was left sitting and decaying in the fields because sale prices could not cover the production costs. Nutritionists estimate that

Table 6.3   Food waste in Italy

Food type	Available quantities (g/person per day)	Waste (%)
Cereals, cereal products and substitutes	154.59	36
Vegetables, fresh and processed	233.34	48
Fruits, fresh and processed	140.54	41
Alcoholic beverages and substitutes	97.1	44
Meat, meat products and substitutes	98.51	42
Fish and seafood	15.42	26

Source: authors 'elaboration based on FAO and INRAN data.

the average consumption of fruits and vegetables per year should be 182.5 kg, so the quantity of fruits and vegetables wasted could have allowed more than 48 million people to consume the proper amount of fresh fruits and vegetables for a year.

- **Waste at producers' organizations level**: in 2008/09 almost 81 000 t of fruits and vegetables were withdrawn from the market, of which only 6.13% was not wasted.
- **Waste at agri-food-industry level**: in 2009 food waste at this level stood at around 2.2% (2 161 312 t). More than 2 Mt of meat, beverages and dairy products are thrown away every year—enough to feed the entire Veneto region. Wasting 250 kt of meat (representing 9% of the total Italian food waste) implies wasting 105 ML of water.
- **Waste at retailer level**: in 2009 food waste was 1.2% (244 252 t). Each of the 600 Italian hypermarkets throws away 250 kg of food daily for aesthetic reasons or because it is close to its expiry date. In large and small retail outlets every year about 1.2% of fruits and vegetables (109 617 t) goes to waste.
- **Waste at household level**: it has been estimated that in 2010 Italians threw away 27 kg of still edible food *per capita* with a financial loss of about €454 per family. This figure is positive, considering that the percentage of waste has declined by 13.4% with respect to 2009.

It is worthy underlining that the disposal of food waste is expensive. The food industry alone spends up to 4% of its revenues to dispose of surpluses. The price list of companies specializing in food waste treatment offers collection services at prices ranging from 6 to 80 cents per kilogram depending on the product. It is also important to add that the food wasted could have provided three meals a day for 600 000 people.

Every year Italians waste an amount of food that could meet the food requirements of 3/4 of the Italian population (44 472 914 people). In Italy 13.6 million people live on less than €1300 per month.[15]

6.3   THE REASONS BEHIND NATIONAL FOOD WASTE

Italy, like any other industrialized country, demonstrates inefficiencies in its food system. Food surplus is generated at each stage of the supply chain: at the production, processing, wholesale distribution, retail and household consumption level.

A part from the common reasons leading to food waste (over-production above all), other specific factors can be identified as follows:

- Less than 20% of household income is generally spent on food.
- The cost of handling surplus and unsold products by the food industry is considered as part of routine operational costs.
- Agriculture accounts for only 1.6% of the national GDP.[16]

It is also believed that most of the waste is due to unclear food labelling. In Italy two types of labels are used to set a deadline: a peremptory 'use by', designed for highly perishable products like fresh milk, meat, eggs and fish, and a more flexible 'best before', applied mainly to pasta, yogurt, juices and oil, which is technically the 'date of minimum durability' that indicates the expected date on which the article starts to lose its characteristics without necessarily being harmful to health.

Instead of bringing clarity, this kaleidoscope of definitions actually confuses consumers. More than one-third of people believe that any product past its best-before date should not be eaten, and 53% never eat fruits or vegetables after they have reached that date. Clear food labelling becomes a priority if we consider that every day, on average three times a day, everyone relates to food—*i.e.* to one or more of the 50 000 food products available on the market (from 1974 to 2010 the number of products on sale increased fivefold[17]).

6.4   FOOD WASTE DISPOSAL

Waste disposal in Italy, as in southern Europe generally, still relies largely on landfill. The latest statistics[18] show that landfill is still the most commonly used way to dispose of waste.

In 1997 source separation and material recycling averaged 9.5% of the national municipal solid waste (MSW) production. For final disposal, the figures were:

- Land filling for untreated MSW: 79.9% of total MSW
- Biological treatment (mixed MSW composting): 9.4%

- Incineration (mostly without thermal recovery): 6.6%
- Others (inertization, mixed MSW sorting, *etc.*): 4.1%

The National Waste Management Law (February 1997) also called Decree 22/97 or Ronchi Decree,[19] which also enforced EU Directives 91/156, 91/689 and 94/62, has totally rewritten the Italian legislation on waste. The Decree clearly points out that:

... Land filling comes last in the waste management hierarchy. Waste minimisation is to be preferred along with recycling; then comes thermal recovery and, as a final option, land filling. Only pre-treated waste will be allowed to be land filled since July 2001 and land filling costs and tipping fees will have to take into account savings needed to ensure a long after care period (30 years in latest draft technical regulations, consistent with provisions of the Land filling Directive).[20]

The Ronchi Decree originally set a recycling goal at 35%, to be met by 2009. Now the recycling goal is at 65% by 2012. Source separation of the organic waste is not compulsory, and it is just depicted as a priority. In order to achieve the recycling targets, source separation in Italy is now undergoing an impressive growth. Attention is focused particularly on the predominant waste fractions such as paper and compostable organic waste. Although source separation of organic waste (kitchen and garden waste) is not compulsory, it is becoming the real backbone of the waste management system, yielding on its own (particularly when operated with door-to-door systems) recycling diversion rates as high as 20–40%.

More than 1500 Italian municipalities have adopted the 'door-to-door' separate collection and are already above 55% of source separation; 20% of Italian towns (corresponding to 300 municipalities)[21] have reached the goal of 80% of source separation.

Door-to-door collection takes place in different types of communities; for example, the province of Turin (with 2.5 million inhabitants) has achieved a separate collection rate of 50%. In only 3 years since the start of door-to-door collection the central area of Turin achieved the impressive rate of 60% separate collection. Other cities such as Trento (population 110 000) or Novara (population 100 000) have also achieved good separate collection.

Salerno, in the south of Italy, has attained the highest percentage recycling rate (72%),[22] at the end of October 2009. Capannori, a small municipality in Tuscany (population 50 000), achieved a separate collection rate of 82% in 2008 thanks to the introduction of a door-to-door separate collection. The cost of moving to a door-to-door system was entirely covered by the savings of recycling 16 000 t of waste instead of sending it to disposal. In total, a saving of €2 348 000 million for 2007 allowed a 20% reduction in municipal taxes.[23]

It is worth mentioning that in 2005 a very interesting movement, the Associazione dei Comuni Virtuosi (Association of Virtuous Municipalities),[24] was established thanks to the initiative of four municipalities from different regions: Mons, Colorno, Vezzano Ligure and Melpignano. Such municipalities pursue a sustainable model based on the respect of the territory, reduction in carbon footprint, and sustainable mobility centred on the principle of car sharing, bike sharing, car pooling, integrated public transport and energy efficiency. The issue of waste disposal and waste recycling is central, as well as the involvement of citizens and the raising of public awareness. Currently there are 45 member towns and every year an annual prize is award, with the aim of rewarding local sustainable practices.

6.5   FOOD RECOVERY PROGRAMS

In Italy, just as in the other European countries discussed in the previous chapter, a number of waste prevention and food recovery initiatives are running at the present time (2011). Most of these initiatives are local, informal, run on a voluntary basis and frequently established by religious entities. Informality seems to be the main characteristic of food rescue initiatives in Italy, whose aim is to recuperate still edible but no longer saleable food and products coming from supermarkets, shops, canteens and restaurants.

The distribution of food for social purposes is governed by Law No. 155, also known as the Good Samaritan Law, which came into force in July 2003: it allows all non-profit organizations working for social solidarity to recover highly perishable foods which remain unsold along the food supply chain and distribute them to people in need. The rationale behind the legislation is to encourage and facilitate the recovery of still perfectly edible food and food products, whose only disadvantage is that they have lost market value and thus would be excluded from the mainstream market. Such a law encourages the self-empowerment of those who decide to get involved in the activity of recovery, and consequently supports poor people whose health is often precarious.

There are many important food rescue initiatives that deserve to be mentioned here, including Remida Food[25] a project undertaken by the municipality of Reggio Emilia, and the Buon Samaritano[26] a scheme launched by Amiat, the urban health company of Turin, which ensure the transportation of rescued foodstuff from hypermarkets and school canteens to soup-kitchens and charities.

Other no-waste initiatives vary according to the stakeholders involved. As in the previous chapter we will try to cite the most important ones according to type.

6.5.1 Restaurants

The first network of 'no waste' restaurants[27] was launched in 2010 in Milan. The 25 restaurants which constitute the network provide their customers with a doggy bag so that they can take leftovers home.

In Piedmont more than 100 bars and restaurants participate in an initiative called *Buta Stupa* (meaning 'uncorked bottle' in Piedmontese dialect); those who don't drink all of their bottles of wine can take them home packed in an elegant bag.

6.5.2 Charity Associations and Non-Profit Organizations

The main non-profit organization that deals with the collection and donation of public and private food surpluses is *Banco Alimentare*, the National Food Bank Foundation.[28] Its headquarters are in Lombardy, Milan and Muggiò (where the regional collection centre is located) but it operates in 19 regions. Based on the concept of giving and sharing, the work of the Food Bank—which has parallels in the United States and elsewhere in Europe—is expressed in the collection of industrial surplus food production (specifically rice, pasta, olive oil, milk). The Food Bank is only rarely able to provide fresh fruits and vegetables. These surpluses are then redistributed to charitable organizations, such as parish counselling centres and addiction rehab associations.

Every year Banco Alimentare organizes the National Food Collection Day. On this day in many supermarkets throughout Italy people can buy food for persons in need: such products are handled to volunteers who collect and donate them to the charities. In 2009, 7600 supermarkets and 100 000 volunteers took part in the National Food Collection Day, allowing the collection of 8976 t of food for an economic value of more than €27 million.

Siticibo[29] is another charity association operating in Milan in 2003 as a sub-program of Banco Alimentare. Siticibo retrieves unsold food from canteens, schools, hospitals, hotels and other catering facilities and donates it to organizations which provide meals to the poor. From unserved cooked food to fruits, from vegetables and bread to desserts, Siticibo vans collect the surplus food and bring it where it is most needed. In 2009 almost 13 000 kg of bread, 9350 kg of fruit and vegetables, 16 490 portions of cooked fresh food and 8500 kg of various crops were donated.

Società del pane quotidiano[30] is a charitable organization based in Milan and helping between 2500 and 3000 people (mostly elderly, unemployed or immigrants) by providing bread, milk, yogurt, fruits, vegetables, chocolate, biscuits and pasta. Such products are donated to Pane Quotidiano directly by food companies. The daily number of visitors has increased from 80–100 in 1970 to 3000 in 2010, for a total of 660 000 people fed during that year.

6.5.3   Retailers

Instead of sending food to landfill, retailers often sell it to stores like *Quel che c'è*[31] which operates in Milan and online; it sells food in stock and provides discounted food.

In 2007 the *Coop* has launched *Buon Fine*,[32] a project with the aim of rescuing food that would be otherwise thrown away. On the national level it involves 380 stores, 1301 non-profit organizations and 123 000 beneficiaries. In 2009 it saved from landfill and distributed more than 2.4 million kg of food worth €14 million.

Esselunga[33] has signed an agreement with the Food Bank and has in 2009 recovered food worth €1 million. *Conad*[34] and other big names are launching similar projects.

Although it is not specifically a food recovery initiative, it is worth mentioning here the *Slow Food* movement[35] founded in Italy in 1986 by Carlo Petrini. This is an eco-gastronomic organization that has grown into an international movement and promotes good, clean and fair food.[36] The fight against food waste is one of its aims alongside animal welfare, consumer health and fair compensation for food producers.

At the core of the Slow Food movement is the belief that consumer's food choices have global consequences. Consumers are thought to be 'co-producers' of their food, because through their food choices they bear some responsibility for how their food is produced. Slow Food trains consumers' senses to rediscover the joys of eating fresh local foods and explains how food production affects people's lives and the environment. The Slow Food movement has three complementary pillars: to educate consumers; safeguard food, local cuisines and traditional products; and connect people across food communities. Petrini has also founded *Terra Madre*, a global network that connects farmers, breeders, fishermen, cooks and agricultural experts for knowledge exchange.

6.6   LAST MINUTE MARKET (LMM)

Type of initiative: **academic spin-off and food recovery initiative**

Main type of stakeholder targeted: **people in need with no or little purchasing power**

Geographic level of implementation: **Italy**

Year of implementation: **1998**

Website: **www.lastminutemarket.it**

Last Minute Market is an unusual food rescue initiative that differs from the projects presented above. It is a very comprehensive action against food waste operating on a national scale in Italy. The initiative is of vital importance to the present authors because it was launched and implemented by one of us, Professor Andrea Segrè, Dean of the Faculty of Agriculture of the University of Bologna, Italy.

Last Minute Market's story began in 1998 when Andrea Segrè—at that time Professor in Agricultural Politics at the Faculty of

Agriculture of the University of Bologna—organized a visit to a local supermarket for a group of students. The aim of the visit was to explain the functioning of the supply chain and the marketing strategies used by retailers. Before the end of the visit, the supermarket manager took us to the back of the store where piles of food were lying on the ground. That food would have been thrown away the next day merely for aesthetic reasons. As a consequence of this shocking visit, we started to think how that food could be recovered and perhaps donated to charities, thus creating a link between for-profit organizations (the retailers) and non-profit organizations (the charities).

Our pilot study started as an experiment: the aim was to provide reliable data having as a reference point the available facts and figures of a benchmark mid-size supermarket and a group of school cafeterias. The objective of the analysis was to determine the amount of calories provided by different types of foods recovered in an average supermarket in the time span of a month, and subsequently extrapolate these data to the national level. The calories available from rescued foods had then to be correlated with the daily average calories needed per individual, based on FAO and 'livelli di assunzione giornalieri raccomandati di nutrienti per la popolazione italiana' (LARN) data (*e.g.* 3680 calories for an average man doing heavy manual work). Such estimates led to the formulation of a hypothesis concerning the number of people that could have been fed with the recovered food. The overall model was estimated using the OLS method, a statistical technique that investigates the association between dependent and independent variables and determines the line of best fit for a set of observations.

Basically, what we found was that an average supermarket ($6000\text{--}7000\text{ m}^2$) could recover nearly 146 000 t (net weight) of food products every year, which could be used to assist 28 466 people and provide 71 000 meals (190 meals per day). From a nutritional point of view these foods are identical to the products for sale on the supermarket shelves.

The theoretical phase (1998–99) was followed by a year of applied research in 2000 involving a group of local stakeholders in the Bologna area. The Last Minute Market (LMM) prototype was finally established in 2001 and the project officially started in 2003. LMM was first established as a cooperative but in 2008 it became an academic spin-off supported by the University of Bologna.

LMM currently operates in 42 Italian towns and offers services to enterprises and institutions in order to prevent, reduce and recover waste and unsold goods. LMM functions as an intermediary that links supermarkets and shops—the so called suppliers—with charities—the so called demanders. The reduction of food waste is just one of its main outcomes.

LMM's core principles are:

- Close connection to the territory where it operates (it involves municipalities, commercial enterprises, wholesale markets, local non-profit, multi-utilities, local health centres, associations, citizens).
- The environmental sustainability of its actions (food should be collected and consumed within a limited radius, thus reducing pollution).
- Its social value (LMM creates a network of active solidarity between profit and non-profit organizations and aims at strengthening social cohesion).
- The indirect economic advantage it provides (lower storage and disposal costs for retailers and shops in general).

LMM does not directly manage the food recovered but sets up an efficient recovery system which is safe from the administrative, fiscal, nutritional and organizational points of view. LMM offers assistance to retailers and shops in all the recovery phases: from the planning (in order to define how to adapt the model to the specificities of a particular area) to the assessment, management and coordination of a well-organized recovery operation system.

The LMM model can be applied not only to the recovery of food but also to the recovery of other goods that are excluded from commercial channels but are still perfectly usable.

LMM has six different and interrelated areas of activities:[37]

- **LMM Food:** for the recovery of unsold foods which are still edible (its functioning is explained in details in the next section).
- **LMM Harvest:** for the recovery of vegetables not harvested due to market reasons or weather damage. Between July and

September 2010, 18 700 kg of potatoes and 200 kg of melons and watermelons, which would have been abandoned in the fields, were harvested. The recovery was possible thanks to a joint initiative of five farms on the outskirts of Bologna and Modena, four non-profit organizations, Caritas, and two drug rehabilitation centres. The rescued food was distributed to people in need.

- **LMM Seeds:** for the recovery of seeds that do not comply with the market standards. Seeds whose germination rate is slightly slower than the minimum set by the EU regulations cannot be sold on the market. As a consequence they are sent to landfill and burnt, as they potentially present a great risk of biological pollution. LMM Seeds recovers these seeds and ships them to farmers in developing countries. LMM Seeds is the only component of LMM that does not take mileage and proximity into consideration.

- **LMM Catering:** for the recovery of products not served by public and private catering services. In November 2009 LMM, in coordination with Hera and Concerta, two leading catering companies, launched the Ciboamico project. Thanks to this project, 15 000 meals were saved and donated in just one year (November 2009–November 2010).

- **LMM Book:** for the recovery of unsold books that would be otherwise destroyed. Since its launch in 2004, more than 44 000 books have been saved from pulping. They were donated by a number of Italian publishers to foreign institutions (schools and Italian Institutes of Culture in Argentina, Brazil, Uruguay, Cuba), as well as to organizations within Italy (senior citizens' centres, schools, hospitals, prisons, public libraries, non-profit organizations).

- **LMM Pharmacy:** for the recovery of pharmaceuticals (non-prescription medicines, baby products, free samples) that could be of use to socially disadvantaged people. In 2009, in the Veneto region alone, pharmaceuticals valued at €15 000 were collected.

Since its beginnings LMM has been able to cope with many challenges, like a general lack of trust in the likely success and effectiveness of the project, a complicated bureaucracy and the need to prove to the local health authority that the food recovered was still good, healthy and edible, intact in all its nutritious components and therefore good enough to be consumed without health-related risks.

6.6.1 LMM Win–Win Model

LMM is working with major large retailers and dozens of institutions such as municipalities, provinces, regions, local health authorities, environmental services companies, public and private foundations, hospitals, farmers' markets, schools and libraries. The beneficiaries are various organizations offering assistance to the most vulnerable people.

In agreement with local municipalities, LMM dictates the 'rules of the game', *i.e.* the best interpretation and application of regulations to ensure maximum safety, sanitary, fiscal and administrative solutions. Qualitative food security is assured by the adoption of the Hazard Analysis and Critical Control Points (HACCP) protocol.

The project is designed as a service to:

- commercial companies
- charities and other non-profit organizations
- public institutions (municipalities, provinces, regions, local health authorities)
- waste removal companies.

LMM assures benefits to all its stakeholders.[38] In particular, thanks to LMM:

- **Commercial companies**, *i.e.* hypermarkets, supermarkets, local shops, reduce the cost of landfill and obtain fiscal advantages: retailers deduct the VAT from the cost of the food they donate, and because of the new Good Samaritan Law, delegate any responsibility concerning food quality and conditions to LMM. As a result, commercial companies can save money, optimize their logistics and pay for LMM's service, which performs the essential task of sorting and separating the edible but unsold foods from the inedible ones. Furthermore, by participating in an ethical and sustainable initiative like LMM, commercial companies can also improve their public image.

- **Public authorities** (municipalities, provinces, regions, local health authorities) and *waste removal companies* reduce the quantities of products thrown away, improve the quality of the assistance provided to people in need, activate a dynamic and stable network between profit and non-profit organizations and guarantee that the rescued goods are used for social purposes.
- **Charitable organizations** receive a wide array of products free of charge and so have more money to allocate for other types of goods and services, thus improving the quality of the assistance provided.

Table 6.4 illustrates the benefits for all the participating LMM stakeholders.

Furthermore, LMM provides benefits for society as a whole by contributing to the promotion of ethical behaviour and by enhancing the importance of reciprocity. More food on the plates of the poor means less food in landfill, less pollution, increased social welfare, poverty reduction and a consistent decrease in food waste. At hypermarkets and supermarkets where the project runs it has been noted that the quantity of discarded products gradually diminishes.

6.6.2   How LMM Works in Detail

The first phase of LMM activities starts in the hypermarket/supermarket where a specific area is reserved for the storage of products that, in commercial terms, can be no longer sold but are still edible (Figure 6.1). Supermarket staff, trained by LMM, sort and check the products: boxes of pasta, soft drinks, baby food, canned tuna, packages of snack food and much more are analysed one by one (Figure 6.2). Expiry dates are checked on each package: the integrity of products intended for human consumption is double-checked. Food that is designated for animal consumption (like pasta in damaged or broken packages) is subject to less strict control.

One by one, all the recovered products are catalogued and 'devalued', *i.e.* removed from the inventory of the supermarket, so that they can easily be donated to charities and associations. Once this is done, the staff prepare the boxes for distribution. Just the

Table 6.4 Benefits for the stakeholders involved in LMM

Stakeholders	Benefits	Costs
Charities and non-profit institutions	They have a constant and free food supply They can provide better assistance to people in need They can reinvest the savings generated from goods received for free	They have to pay for food transportation
Commercial companies	They reduce disposal costs They decrease the rate of unsold products They contribute to corporate social responsibility They obtain fiscal benefits They increase their visibility in the territory	They undergo minor changes in internal management
Local government	They optimize resources They reduce the environmental impact of food waste They participate in a solidarity network They contribute to a general improvement in the health conditions of people assisted	They bear LMM's activation cost They may have to coordinate other local stakeholders

Source: LMM data elaboration.

right amount of food is prepared for the recipient charities (Figure 6.3), based on the number of people they serve. Unfortunately it is not always possible to predict what types of food will be donated by the supermarket, but LMM does its best to assure a wide variety of food.

At the end of the day, it has been estimated that each charity receives food worth €400–1000. Considering that each charity is allowed to collect food two or three times per week, it is evident that LMM allows them to spare a significant sum of money every month.

LMM is designed to optimize time and resources (Figure 6.4). The rapidity with which fresh products are consumed is a crucial point, because the transportation of fresh food and the expiry date involve risks that cannot be underestimated. Once the food

Figure 6.1   LMM staff at work in the back yard of a supermarket.

Figure 6.2   LMM—food recovery.

is donated, it must be preserved and cooked properly. LMM staff assume the responsibility of the good quality of the food recovered, while the charities assume responsibility for the good handling of it.

Figure 6.3 Beneficiaries of LMM: the charities receive and collect free food daily.

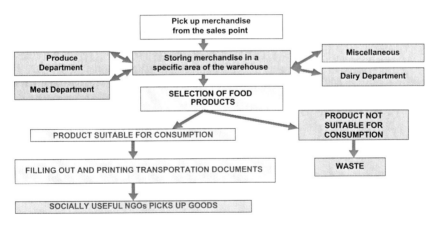

Figure 6.4 How LMM works. Source: LMM data elaboration.

The rules are clear, and non-profit organizations are required to observe them. Periodically, a member of LMM staff pays a visit to the charities and makes sure that the food is properly stored and cooked in the charities' kitchens (Figure 6.5).

Figure 6.5 A charity kitchen.

6.6.3 What Can be Recovered

The many food products that can be recovered include the following:

Fruits and Vegetables. The greengrocery department accounts for 65% of LMM's recovered products. Boxes of oranges and apples, packets of ready-made salad, and much more are rescued every day.

Meat. Up to 18% of LMM products can be recovered from the butchery department. Only packed meat can be donated to charities: it is normally withdrawn from the supermarket shelves 3 days before the expiry date.

Canned and Ready-Made Foods. Cans of tuna, beans, peas, fruits juice, biscuits, snacks, baby food, pasta can be recovered. In most cases these products are withdrawn because their packaging is damaged or broken. In addition to long shelf life products, cheese and yogurt, fresh pastas, gnocchi, butter and all refrigerated products can be collected by LMM. The supermarkets withdraw them from 2 to 6 days before their expiry date.

Bakery. Pizza slices, pastries, cakes and cookies are removed every day from the supermarket shelves and donated on the very same day: the amount of food recovered in this case is not very high, as supermarkets try to calculate the right amount needed in order not to have too much surplus. LMM prefers to avoid the collection of pastries filled with cream or whipped cream. Although prepared in the premises of the supermarket and removed from the shelves 24 hours after their production, such confectionery can deteriorate easily.

Frozen Foods. These are recovered in small quantities because they need to be stored and transported with extreme care in polystyrene boxes or insulated bags.

Fish. Fish is not recovered as it is considered too delicate to be handled. The rapid deterioration of fish, especially during hot summer months, has prompted LMM to exclude it from the list of products recovered.

CASE STUDY: Food Recovery from School Cafeterias in Verona

One of the most representative case studies is that of Verona, a city in the north of Italy, where an efficient food recovery scheme from school canteens and cafeterias has been established.

The *Cafeterias beyond School* project started in 2007 and now involves seven educational institutions and many other stakeholders such as the Italian Christian Workers Association of Verona (as promoter, financer and coordinator for LMM), the School Board of the Municipality of Verona (as promoter of all activities related to the collection of unserved meals from school cafeterias), the Local Socio-Sanitary Unit 20 Verona (as promoter and supervisor for health-related issues), Amia Verona (the local environmental health company), the Vincenziano volunteer group and a social cooperative (as beneficiaries of the products collected from the school cafeterias).[39]

The project is designed such that cooked unserved meals can be collected immediately after the cafeterias close. The advantage is that the recipient associations receive a ready-to-eat meal with high nutritional value. The disadvantage is that these cooked meals must be transported in special insulated boxes for their entire journey and the temperature must be maintained and checked very frequently (it should never go below 65 °C). Table 6.5 shows the amount of food recovered in 2007.

On average, 18% of the meals are recovered every day. The surplus is not due to the poor management of the meal preparation department but rather to the fact that portion sizes are sometimes too big and food does not always correspond to the tastes and appetites of the children.

Thanks to LMM, the Municipality of Verona has increased the efficiency and effectiveness of its services. The achievements can be summed up as follows:

- Consistent reductions in the costs of waste removal
- Daily monitoring of food products
- Optimization of orders and purchases
- Specific menu design
- Analysis of children's eating habits
- Increased parental awareness of the nutritional aspects of food
- Increased public awareness of food
- Increased positive media coverage of the activities of the Municipality of Verona
- Improved services for people in need.

Table 6.5   The collection at five school cafeterias in the Municipality of Verona in 2007

Products	Quantity (kg)	%
Pasta	2533.60	38.84%
Main course	960.84	14.73%
Side dish	1883.06	28.87%
Fruit/yoghurt	144.26	2.21%
Other	101.01	1.55%
Bread	900.58	13.81%
Total	6523.35	100.00%

Source: LMM data elaboration.

6.6.4 If LMM were to be adopted all over Italy ...

In 2008, from supermarkets alone, nearly 170 t of good edible food (made up for 60% of fruits and vegetables, 9% meat, 12% canned and other packaged products, 6% milk, 13% bread and pastry) were recovered, with a value of €646 000. With this food it was possible to provide 365 000 meals and guarantee assistance for 400 people per day.

In 2009 34 000 kg of food was recovered from small independent shops and consequently it was possible to provide 131 000 meals and to guarantee assistance for 100 people per day.[40]

Another example of the positive effects induced by LMM is that of Sardinia where, between March and December 2006, 31 056 kg of products (726 kg of beverages, 13 700 kg of fruits and vegetables, 4000 kg of dairy products, 2130 kg of meat and eggs, 3500 kg of canned food and 7000 kg of bread) were recovered from two supermarkets (Conad and Carrefour). The beneficiaries were two Caritas canteens.

Table 6.6 shows the estimated quantity of food and other products recuperated thanks to LMM in a sample of some representative Italian towns in 2009.

Quantitative and qualitative data analysis have proved that LMM brings about environmental, economic and social benefits (Table 6.7). If LMM Food were to be adopted nationwide by supermarkets, small shops and cash and carry outlets, it would be possible to recover food worth €928 157 600. Three meals a day could be provided to 636 000 people, for a total of 580 402 025 meals in a year. It also important to underline that, by avoiding sending food to landfill, 291 393 t of CO_2 emissions could be

Table 6.6 Quantities of products recovered per year by LMM according to city and type of store

City	Type of store	Quantities of products recovered
Bologna	1 superstore	170 t/year (worth €750 000)
Ferrara	15 small shops	60 t/year
Verona	8 school canteens	15 000 meals/year
Cesena	Fruits and vegetable market	60 t/year
Ferrara	11 pharmacies	€15 000/year
Italy	12 publishers	47 000 books

Source: LMM data elaboration.

Table 6.7   Quantities of food recovered per year by LMM plus economic and environmental impact of LMM

Type of outlet	Expected quantities of recuperated food per year
Cash and carry shops	4644 t
Superstores	67 467 t
Supermarkets	128 785 t
Small shops	43 357 t
Total	244 252 t
Value of recuperated food	€928 157 600 (= 580 402 025 meals) in a year
CO_2 emissions	291.393 t of CO_2 are produced because the food ends up in landfills

Source: LMM data elaboration.

spared. In the case of pharmacies, medicinal products to the value of €597 504 600 could be recovered.

6.6.5   Nutritional Impact of LMM

Based on the amount of food recovered and distributed by a mid- to large-size supermarket and the related calories for each type of food (defined in detail by LARN) the annual amount of calories recovered can be calculated as 149 650 994.17 kcal[41] (Table 6.8).

It is worth noting that only 70% of these calories are intended for human consumption. Considering that the average energy need for a male manual labourer per day has been estimated by FAO at 3680 kcal, it has been calculated that nearly 28 466 people could be helped and 71 000 meals could be provided per year by each supermarket.

The same analysis can be conducted in a different way, dictated by nutritional needs and verified by the empirical nature of the distribution. Since approximately 0.5 kg of food are consumed per meal, it has been estimated that 70% of the 146 t of products recovered by every single supermarket could provide 204 000 meals per year or 560 meals per day. To put it another way, 10 422 216 people could be fed per year and 70 000 meals could be recovered from Italian supermarkets (Table 6.9).

6.6.6   Economic and Environmental Impacts of LMM

As well as its nutritional benefits, LMM has proved to lead to economic benefits, as it contributes to a reduction in the disposal costs for all the commercial companies involved in the initiative,

Table 6.8 Total calories produced and distributed in a year by a medium-sized to large supermarket

	kcal
Dry goods	39 107 977
Dairy	16 309 602
Meat	46 422 354
Other products	34 674 687
Bakery	13 136 372
Total	149 650 994

Source: LMM data elaboration.

Table 6.9 Estimates of the total calories that can be produced and distributed per year from the total number of domestic supermarkets

	10^9 *kcal*
Dry goods	15.40
Diary	6.03
Meat	17.10
Produce	12.87
Bakery	3.27
Total	54.68

Source: LMM data elaboration.

and to contribute to environmental benefits, as it has been calculated that 10% of greenhouse gas emissions of developed countries comes from the production of food that is thrown away.

Food waste prevention and recycling are powerful strategies in reducing greenhouse gas emissions and conserving energy.

- Waste prevention and recycling divert organic wastes from landfills and reduce the amount of methane released when these materials decompose.
- Recycling and waste prevention allow some materials to be diverted from incinerators and reduce greenhouse gas emissions coming from the combustion of waste.
- Waste prevention is effective at saving energy, as less fossil fuel is burned and less carbon dioxide is emitted into the atmosphere.

In order to estimate the amount of food and products generated and recycled (or discarded) at the national level, LMM staff used a materials flow methodology (MFA) which relies heavily on a mass balance approach. The material flow analysis is a quantitative procedure used to calculate the flow of materials and energy through the economy. It uses input/output methodologies, including both material and economic information.[42] It quantifies the linkage between environmental problems and human activities, serves as a systems-wide diagnostic procedure, supports the planning of adequate management measures and monitors the efficacy of those measures. MFA allows early warning and supports precautionary measures.

As far as food recovery is concerned, data gathered from hypermarkets, supermarkets, small businesses, wholesalers, school, barrack and hospital canteens were used. The estimations reported in the box below are therefore based on concrete data gathered on a survey conducted in 2009.

- It has been estimated that 1 t of recovered food is equal to sparing:
  - 1193 kg of CO_2 not released into the environment
  - 4.5 t of CO_2 (if we consider pollution caused by food production, storage and transportation)
- The production of 1 kg of meat uses 50 000–100 000 L of water
- The production of 1 kg of rice use 1500 L of water
- The production of 1 kg of corn uses 540 L of water
- 1 t of recovered food has a value of €3700, and €50–180 is spared thanks to the avoidance of landfill

Source: authors' elaboration

The data reported in the following tables provide a national picture of food waste generation, management and recovery correlated with an estimation of how many people could be fed with the rescued food. It is worth noting that there are many regional variations, depending on the local and regional availability of suitable landfill space, the proximity of alternative markets for the recovered materials and the population density.

6.6.7 A Year against Food Waste and the Joint Declaration against Food Waste

In 2010 LMM launched a project entitled 'A Year Against Waste' supported by the Faculty of Agriculture of the University of Bologna, in coordination with the Technical and Scientific Conference of the Deans of the Italian Faculties of Agriculture and under the patronage of the Committee on Agriculture and Rural Development of the European Parliament.

The project aimed to raise public awareness of the causes and consequences of waste in Italy and Europe, provide strategies on how to reduce it and promote a culture oriented to the principles of sustainability. Several events formed part of the project: the launch of the *Black Book of Italian Waste* written by researchers of the Faculty of Agriculture of Bologna, the organization of a free lunch in the main square in Bologna where more than 1000 people were given food that would have been otherwise thrown away (Figures 6.6–6.8), a conference at the European Parliament in Brussels entitled 'Transforming Food Waste into a Resource' (Figure 6.9), an award intended to reward the most environmental friendly company, and a comedy piece for theatre.[44]

The conference in Brussels was one of the highlights of the project. Conference participants included Paolo De Castro (Chairman of the Committee on Agriculture and Rural Development of the European Parliament), members of the European Parliament, politicians, academic professors such as Paul Connett

Figure 6.6   Free lunch for 1000 people organized in Bologna by LMM on 30 October 2010, as part of the project 'A year against food waste'.

Figure 6.7   Professor Andrea Segrè (on the left), at the free lunch in October 2010.

(St. Lawrence University, USA) and Jan Lundqvist (Stockholm International Water Institute, Sweden), NGOs and organizations like the Stop Wasting Food movement (Denmark), A.N.D.E.S. (France) and FareShare (England). At the end of the conference, a Joint Declaration Against Food Waste was announced and presented to the members of the European Parliament. This document contains proposals for sustainable use of food and commitment to the global reduction of food waste by at least 50% by 2025 through the involvement of all stakeholders from farm to fork, *i.e.* farmers, distribution, marketing system and consumers. It advocates the creation of a Global Partnership against Food Waste that involves a large number of stakeholders and community members and commits itself to the sharing of technologies, processes, projects and ideas for increasing the capacity of global, European, national and regional institutions and governments to find solutions to food waste (Figure 6.10).

The Joint Declaration also calls on the United Nations to include the fight against food waste as an additional target within Goal 7 ('Ensure environmental sustainability') of the Millennium Development Goals[45] and urges a coordinated and agreed way to

Figure 6.8   The square in Bologna where the free lunch took place.

Figure 6.9   How to Transform Food Waste into a Resource—Conference at the European Parliament (28 October 2010). From left to right: Professor A. Segrè, Professor P. Connett and Professor J. Lundqvist.

Figure 6.10 Professor A. Segrè and Vandana Shiva at a conference in the Salone del Gusto, Turin.

reduce food waste. It urges national governments and organizations such as the Food Standards Agency to develop practical solutions (*i.e.* the promotion of transparency in labels as well as more appropriate packaging solutions) and requires that the fight against food waste becomes one of the priorities on the European Commission agenda.

CONCLUSION

The LMM philosophy is to transform waste into resource. LMM is apparently so simple as to seem trivial. It consists in fact in recovering what is still useful and giving it to those in need, thus contributing to less waste, less pollution, more sustainability, more savings, more investments, more solidarity and, last but not least, improved health conditions for those who benefit from its service.

LMM constitutes an important discovery: without its daily operation larger quantities of food and other products would be wasted. It differs from all other previous recovery programmes, both national and international, because it is the only 360° action against waste as it deals with food (including fresh goods like fruits and vegetable) and non-food products.

LMM is about combining logistics with solidarity through sustainable actions. LMM also contributes to positive and constructive relations between people. Some scholars, like the Italian Maurizio Pallante[46] consider LMM as an example of economic degrowing, sobriety and energy sparing.

In the last years (2005–2011) LMM has featured in many Italian and international scientific papers, magazines, newspapers and TV programmes. It's important to underline that LMM is not protected by copyright. LMM staff believe in an open society where progress is brought about by ideas—as in the case of LMM—which can be copied, implemented and become useful concrete projects with the aim of making the world a better and more sustainable place.

APPENDIX: LAST MINUTE MARKET FACTS AND FIGURES[43]

Estimated impacts of LMM FOOD

Table 6.10 Recovery at an average hypermarket

Quantity of food recovered	170 t
Economic value	€646 000
Number of people given 3 meals per day	371
Meals per year	338 538

Source: LMM data elaboration.
  If this food had not been recovered, 203 t of CO_2 would have been produced. To neutralize it, a wooded area of 408 ha, equal to 816 football fields, would have been needed.

Table 6.11 Recovery at an average supermarket

Quantity of food recovered	45 t
Economic value	€171 000
Number of people given 3 meals per day	99
Meals per year	90 338

Source: LMM data elaboration.
  If this food had not been recovered, 54 t of CO_2 would have been produced. To neutralize it a wooded area of 109 ha, the equivalent of 218 football fields, would have been needed.

Table 6.12 Recovery at an average fruit and vegetable wholesale outlet

Quantity of food recovered	31 311 kg
Economic value	€75 150
Number of people given 3 meals per day	172
Meals per year	125 560

Source: LMM data elaboration.
If this food had not been recovered, 36 t of CO_2 would have been produced. To neutralize it a wooded area of 77.5 ha, the equivalent of 151 football fields, would have been needed.

Table 6.13 Recovery at an average small shop

Quantity of food recovered	49 t
Economic value	€186 200
Number of people given 3 meals per day	107
Meals per year	97 640

Source: LMM data elaboration.
If this food had not been recovered, 56 t of CO_2 would have been produced. To neutralize it a wooded area of 118 ha, the equivalent of 236 football fields, would have been needed.

Estimated impacts of Last Minute Market CATERING

Table 6.14 Recovery at eight school cafeterias in Verona

Quantity of food recovered	10 170 kg
Economic value	€38 650
Number of people given 3 meals per day	22
Meals per year	20 075

Source: LMM data elaboration.
If this food had not been recovered, 12 t of CO_2 would have been produced. To neutralize it a wooded area of 24.5 ha, the equivalent of 49 football fields, would have been needed.
Last Minute Market CATERING recovers food from hospitals and barracks, too.

Table 6.15 Recovery from two hospital canteens

Quantity of food recovered	16 425 kg
Economic value	€62 415
Number of people given 3 meals per day	36
Meals per year	32 485

Source: LMM data elaboration.
If this food had not been recovered, 19 t of CO_2 would have been produced. To neutralize it a wooded area of 40 ha, the equivalent of 80 football fields, would have been needed.

Table 6.16   Recovery from an average military barracks

Quantity of food recovered	3859 kg
Economic value	€14 664
Number of people given 3 meals per day	10
Meals per year	9125

Source: LMM data elaboration.
  If this food had not been recovered, 4.5 t of CO_2 would have been produced. To neutralize it a wooded area of 9 ha, the equivalent of 18 football fields, would have been needed.

Estimated impacts of Last Minute Market HARVEST

Table 6.17   Recovery at a farmers' association

Quantity of fruits and vegetables recovered	18 900 kg
Economic value	€47 250
Number of people given 3 meals per day	104
Meals per year	75 920

Source: LMM data elaboration. If this food had not been recovered, 22 t of CO_2 would have been produced.
  To neutralize it a wooded area of 45.5 ha, the equivalent of 91 football fields, would have been needed.

Estimated impacts of Last Minute Market SEED

Table 6.18   Recovery from a seed industry

Quantity of seeds recovered	5000 kg
Economic value	€18 900
Number of people given 3 meals per day	6500

Source: LMM data elaboration.
  If the food had not been recovered, 6 t of CO_2 would have been produced. To neutralize it a wooded area of 12.5 ha, the equivalent of 25 football fields, would have been needed.

REFERENCES

1. L. Fulponi, *Changing Food Lifestyles: Emerging Consumer Concerns*, AgrirRegioniEuropa, anno 1, n. 3, December 2005.
2. http://ec.europa.eu/food/food/labellingnutrition/foodlabelling/fl_com9138_annex_i_en.pdf
3. A. Carbone and M. De Benedictis, *Trasformazione e Competitività del Sistema Agroalimentare Italiano nell'UE Allargata*,

Economia italiana, Rivista quadrimestrale, n. 1, January - April, Feltrinelli, 2010.

4. A. Segrè and A. Grossi, *Dalla fame alla Sazietà*, Sellerio Editore, Italy, 2007, p. 181.

5. Istat, *Rapporto Annuale, La situazione del Paese nel 2005*, Capitolo 5, Disuguaglianze, Disagio e Mobilità Sociale, Roma 2005.

6. http://www.uniconsum.it/il-consumatore-e-i-suoi-diritti/1585-la-spesa-mensile-delle-famiglie-italiane.html

7. Italian Ministry of Public Health Reports (2002, 2008).

8. http://www.fao.org/ag/agn/nutrition/Indicatorsfiles/FoodSupply.pdf

9. ISMEA, *La Competitività dell'Agroalimentare Italiano, Check-up 2005*, Redazione a cura della Direzione Mercati e Risk Management, 2009.

10. http://www.andreasegre.it/carocibo

11. http://www.repubblica.it/2009/10/sezioni/economia/poverta-alimentare/poverta-alimentare/poverta-alimentare.html

12. A. Segrè, *Il Libro Nero dello Spreco Alimentare*, Edizioni Ambiente, Italy, 2011.

13. http://www.inran.it/646/tabelle_di_composizione_degli_alimenti.html

14. http://www.cia.it/

15. A. Segrè, *Il Libro Nero dello Spreco Alimentare*, Edizioni Ambiente, Italy, 2011.

16. http://www.agrinews.info

17. E. Peta, *Consumi Agro-Alimentari in Italia e Nuove Tecnologie*, Ministero dello Sviluppo Economico, 2009.

18. http://extranet.regione.piemonte.it/ambiente/aria/dwd/emissioni/lin_guid_inv_loc_rtctn_ace_3_2001.pdf

19. http://www.politichecomunitarie.it/comunicazione/16945/il-decreto-ronchi-e-legge

20. http://www.compostnetwork.info/italy.html

21. http://www.tno.it/tecno_it/indici_it/.../Relazione%20Paul%20Connet.pdf

22. http://www.comune.salerno.it/client/scheda_news.aspx?news=1970&prov=76&stile=7

23. http://www.comune.capannori.lu.it

24. http://www.comunivirtuosi.org

25. http://www.provincia.re.it

26. http://www2.a21italy.it/rifiuti21network/gdl/programma_di_lavoro/torino_19–05–08/Amiat_Torino.pdf
27. http://www.ilsole24ore.com/art/finanza-e-mercati/2011–02–14/anche-ristorante-supermercato-salvano-063901.shtml?uuid = AafBn67C&fromSearch
28. http://www.bancoalimentare.it
29. http://www.siticibo.it/stampa/110103corriersera.pdf
30. http://www.panequotidiano.org
31. http://www.quelchece.it
32. http://www.ecoop.it/portalWeb/portale/common/documento.jsp?cm_path = /CoopRepository/COOP/CoopLombardia/documento/doc00000069004&from_home_page = yes
33. http://www.esselunga.it
34. http://www.conad.it
35. http://www.slowfood.it
36. http://www.ashoka.org/cpetrini
37. A. Segrè, *Last Minute Market. La Banalità del Bene e altre Storie contro lo Spreco*, Pendragon, Italy, 2010.
38. A. Segrè, *Dalla Fame alla Sazietà, dalle Eccedenze allo Spreco Inutile*, Georgofili – Conference Proceedings-Accademia dei Georgofili. Fame e spreco alimentare. Trasformare le eccedenze in risorse a fini solidali. Il caso Last Minute Market. Firenze, Accademia dei Georgofili, 16 October 2006, vol. 3, pp. 455–461.
39. A. Segrè, *Last Minute Market. La banalità del Bene e altre Storie contro lo Spreco*, Pendragon, Italy, 2010.
40. A. Segrè, *Last Minute Market. La Banalità del Bene e altre Storie contro lo Spreco*, Pendragon, Italy, 2010.
41. A. Segrè, L. Falasconi, E. Morganti, *Last Minute Market. Increasing the Economic, Social and Environmental Value of Unsold Products in the Food Chain,* in Waldrom K., Moates G. K., Faulds C. B., Total Food. Sustainable of Agri-Food Chain, RSC Publishing, Cambridge, 2010, pp. 162–167.
42. http://www.zerowasteamerica.org/EconomicsOfWaste.htm
43. A. Segrè, *Il Libro Nero dello Spreco Alimentare*, Edizioni Ambiente, Italy, 2011.
44. http://www.lastminutemarket.it
45. http://www.un.org/millenniumgoals/
46. http://decrescitafelice.it/

CHAPTER 7

Epilogue—or Ode to Sufficiency, Transparency and Efficiency

In this book we have made an attempt to provide a clear picture of current global food policies (with a focus on Europe and the US), of the future challenges connected to food (over)production and (over)consumption, and of the close relationship between food policies and food waste. The main challenge we have faced has been the lack of reliable, high-quality and comparable data about food waste.

Food waste is a much-debated issue but at the same time a contested terrain. The culture of excess we are living in—a 'more is better' culture as Michael Grunwald defines it[1]—works around productive but unfair mechanism where consumers, mainly in the Western world, are used to consider resources as infinite and therefore exploitable.

The hyperconsumption of products has been fuelled through marketing strategies which have supported a culture of throwaway and immediate gratification. The materialistic urge through advertising and marketing has been so strong as to deeply engrain itself in our subconscious. We are taught from a young age to want to shop, to buy or to consume. So 'shop, buy, consume, throw away' are now part of our everyday cultural norms. We can buy now, pay later (and waste in between). In the same way, the earth is heading for a consume now, pay later cross-roads—a point at

Transforming Food Waste into a Resource
By Andrea Segrè and Silvia Gaiani
© Andrea Segrè and Silvia Gaiani, 2012
Published by the Royal Society of Chemistry, www.rsc.org

which we will be forced to consider what we consume now through the consequences of what will happen in the future.

We should shift the focus from quantity to quality, from induced needs to real necessities by keeping in mind that each of our actions—and of our food choices- has environmental, economic and nutritional impacts. We must seek for sustainable processes, as well as products, so to affirm a new logic that focuses on sufficiency and efficiency, intended as a function of quality.

The path of eco-efficiency is a promising and charming one. Daniel Goleman encourages us to imagine our planet as our home and to develop an ecological intelligence, defined as the ability to learn from experience and interact effectively with our environment by reducing our environmental impacts on the resources.[2]

Since we have no innate brain system designed to warn us of the countless ways in which human activity corrodes the planet, we have to develop a new understanding (or in this case intelligence) that allows us to comprehend systems in all their complexity, as well as the interplay between the natural world and human activities. But developing that understanding demands a vast store of knowledge, one so huge that no single brain can store it all. Each one of us needs the help of others to navigate the complexities of ecological intelligence. We need to collaborate and to survive as a collective.

We need to focus on actions and pragmatism, as Anthony Giddens[3] affirms, and enhance our ability to develop smart, ecological solutions that do not focus only on the political rhetoric of change. New business models must emerge that speak to and address 'sustainable' as a 'no waste, no impact' model. Could we imagine Coke Zero being less about 'no' sugar and more about 'no impact' or zero impact on land, water, health and environment?[4]

We have to abandon the principle of efficiency, at least as understood by classical economic theory, in favour of a principle of sufficiency. The principle of sufficiency includes all the others in a comprehensive way; it implies the abandonment of the existing principles of the organization of economic activities and excludes the continuing exploitation of environmental resources. Environmental issues require an interdisciplinary approach that embraces the natural sciences and social sciences and reduces the knowledge fragmentation.

To save us from the 'tyranny of the ephemeral' we need to reinvent our lives around two new principles: lightness and transparency. Lightness refers to sobriety and the abolition of the superfluous. Transparency is a necessity and the result of a choice: we need transparency in communication (consumers need to know how much food is thrown away every day in their country and in the world) and transparency in food labelling. The market should be transparent in order to allow consumers to express their 'vote' by shopping. Daniel Goleman defines radical transparency as the attitude to do good to the economy, the environment and the people.

Food waste is a market and policy failure, and is against the common good. If it cannot be totally avoided, it should be at least reduced or reused. When not suitable for human or animal consumption, food waste can be converted into energy and transformed into gas and liquid fuels, thus providing an alternative to traditional energy which is dependent on non-renewable sources such as oil, coal or uranium. Some food waste can also be used in the production of environmentally friendly products such as biodegradable plastics. In addition, the packaging used for wrapping food can easily be recycled to reduce the amount of waste being dumped in landfills.[5]

As a consequence, waste management experts are now beginning to think that the present system of waste management may have fundamental flaws, and that a thoroughly effective system may need an entirely new way of looking at waste. Talking of 'resource management' rather than 'waste management' is becoming increasingly common.[6] We should stop thinking of waste as a 'waste' and to see it instead as a valuable resource for society. In this sense waste ceases to exist.[7]

To consider food waste as an opportunity is the fundamental concept at the basis of our book, and the pillar of the many food recovery programs we have cited. Last Minute Market, for example, is a 360° action against waste and a successful example of how—if all the stakeholders involved have directly or indirectly benefits and the legislation is properly designed—food (and nonfood) recovery can bring about positive, comprehensive benefits.

In this concluding chapter we have tried to develop some recommendations which are by no means exhaustive, but set out some of the ways towards more sustainable resource management practices and the correct balance of economic, social, political and technological measures.

7.1 ECONOMIC MEASURES

7.1.1 Subsidies and Eco-Taxation

A key means to achieving sound resource and waste management practices is to send effective pricing signals that encourage more sustainable actions.

Although it is difficult to derive an exact value, it is estimated that global subsidies amount to over US$ trillion per year, with OECD members accounting for three-quarters of the total.[8] Phasing out destructive subsidies and shifting a proportion of the funds to resource efficiency initiatives would help address unsustainable resource consumption practices.

Ecological tax reform is the process whereby market prices are adjusted to reflect the full environmental costs of economic activities. Examples include levies on the use of virgin materials, landfill fees, and other waste and pollution charges that incentivize manufacturers to reduce their generation of wastes and emissions.

7.1.2 Green Procurement

Through the products and services they buy, governmental institutions can have a powerful influence over their suppliers. Through the placement of environmental demands, institutions can shift markets and influence design, efficiency and durability. This is 'green procurement'.

Denmark is a world leader in green procurement, with a law in place since 1994 requiring all national and local authorities to use recycled or recyclable products.

Another example of incorporating the cost of environmental impact into commodity prices lies in beverage containers. Sweden has achieved an 86% recovery rate for these, driven primarily through an industry-imposed bottle deposit of 10.[9]

7.1.3 Pay As You Throw (PAYT)

Charging households for the amount of non-recyclable waste they generate has been a successful way to both increase recycling and reduce the absolute volume of waste generated by a population. The first community to implement PAYT was Richmond, California, in 1916, and since then, over 6000 communities in the US alone have implemented PAYT schemes.[10] PAYT charges residents for collection of waste based on the amount they produce, providing a direct economic incentive to generate less waste and increase composting and recycling.

7.1.4   Tradeable Permits

Governments can use permits to regulate and reduce resource consumption, waste generation, and environmental pollution over time. The concept behind tradable permits is that all polluters are sold a defined amount of permits allowing them to discharge waste or pollution. If organizations can reduce their discharge, they can also sell their remaining permits. For example, landfill allowance trading schemes (LATS) allocate tradable landfill allowances to each waste disposal authority in England, allowing the disposal of a certain amount of biodegradable waste per year. These authorities can then use their allocations in the most effective way, such as trading with other disposal authorities or saving them for future years.

7.1.5   Product Service Systems

A whole new way of thinking about products, the way an economy functions, and what it is supposed to accomplish, has recently emerged. Instead of just selling goods, manufacturers are moving towards the provision of services driven by a transformation of consumer habits. In this, customers do not demand products *per se*, but rather seek the utility that products and services provide. By using a service that meets need rather than a physical object, more needs can be met with lower material and energy requirements. Under such a system the emphasis is on quality retail, advising customers on the best leasing option available, on the quality and upkeep of the products, and on how to extend usefulness while using the least energy and materials.

7.2   SOCIAL MEASURES

7.2.1   Influencing Consumption

The global consumer class is key in reshaping patterns of resource consumption, simply because it consumes the bulk of the world's resources. Cleaner technologies and more efficient products and production systems will help reduce the impacts of consumption but, essentially, consuming better does not alleviate the need to consider moderation in overall consumption levels.

7.2.2 Developing a Recycling Culture

Material consumption is used by many people to create and maintain a sense of identity and to show allegiance with certain social groups. Communication and education will therefore have to play a major role in achieving sustainable consumption. People will change behaviour if they understand the reasons for doing so and it is made easy for them. They need to be informed on the environmental and resource-related consequences of their purchasing and lifestyle decisions. Education is also needed to encourage the use of products made from recycled or recovered materials as well as to inform individuals about the importance of source separation of their household waste.

In Germany, schoolchildren are taught about the importance of properly separating their waste, and separate bins are provided and weighed. The less mixed waste you have, the less you pay. In Vienna airport, for example, all public bins have four different-sized compartments: paper, glass, metal and 'other'.

The town of Kamikatsu, Japan, has adopted a goal of zero waste to landfill or incineration by 2020, due to the closing of both local incinerators.[11]

7.2.3 Corporate Social Responsibility (CSR)

Many companies are beginning to address environmental issues because of the positive association such moves will have with their brand. Wal-Mart has been pioneering in its approach to addressing sustainability through several initiatives, including one that concerns packaging. In 2006 Wal-Mart set a target to reduce all supply chain packaging by 5% by 2013.[12]

The sustainable packaging programme is committed to using materials of the highest recycled content without compromising quality. This includes choosing components on the basis of their recyclability after use.

7.3 POLITICAL MEASURES

Governments will play a critical role in moving towards implementing new sustainable waste and resource management processes. Specifically, regulation will allow industry to gain confidence in investing in the development of new technology, as investing in facilities that are not required to meet regulatory standards under legislation is risky.

7.3.1 Product Regulation and Labelling

Governments have the potential to influence product development through regulatory mechanisms. National minimum standards for product performance can be set. Minimum standards are complemented by eco-labelling programmes to give purchasers information and encourage manufacturers to design and market more eco-friendly products.

7.3.2 Integrated Product Policy

At the EU level there is support for integrated product policy (IPP).[13] This aims to influence the environmental impact of products by looking at all phases of their lifecycles—not just the consequences of disposal but also the impacts of production and factors such as energy use in consumption—and taking action where it is most effective. With so many different products, no one simple policy measure can be applied to all. Instead a whole variety of tools, both voluntary and mandatory, can be used to achieve the IPP objective, including economic instruments, substance bans, voluntary agreements, environmental labelling, and product design guidelines.

7.3.3 Extended Producer Responsibility

Waste can be avoided if manufacturers factor in environmental considerations at the product design phase. Extended producer responsibility (EPR) laws encourage this by imposing accountability on manufacturers over the entire product/packaging lifecycle and requiring companies to take back products at the end of their useful life.

EPR typically bans the landfilling and incineration of most products, establishes minimum reuse and recycling requirements, and specifies whether producers are to be individually or collectively responsible for returned products. This mechanism shifts the responsibility for waste from government to private industry and encourages the internalization of waste management costs into product prices. EPR encourages producers to consider waste management and the full lifecycle of the product in the initial product design as it is of advantage for them to do so.

7.3.4 Dematerialization

This process is aimed at reducing the amount of raw materials needed to create a product. Advocates have pushed for a 'factor 10' reduction, *i.e.* policies that aim to provide a given volume of goods and services with one-tenth of the material input.[14] Indeed, there was some success in this area with resource productivity in the EU improving by 30% between 1980 and 1997. However, this improved efficiency has not translated into an overall reduction in resource consumption which instead has remained essentially constant as consumer wants and needs continue to increase. Although dematerialization is an important step towards achieving more sustainable economic activity, in itself it may be insufficient to contend with humanity's increased desire for consumption and thus must be coupled with strategies addressing consumption.

7.4 TECHNOLOGICAL MEASURES

None of today's industrialized economies is truly sustainable and all could be leaner without suffering significant setbacks. Annual material throughputs in America and Europe are estimated as 80 and 51 t per person respectively, but for an average Japanese it is just 21 t.[15] Given the broadly similar living standards of Americans, Europeans and Japanese, clearly there is considerable room for improvement in both the US and Europe.

To conclude, tackling food waste effectively requires a new complexity of governance. This is painful, because short-term measures are required for solutions that will only bring obvious benefits in the long run. The discrepancy between short- and long-term goals affects governments, companies, individuals and institutions. Engaging people in a debate that persuades them to accept short-term sacrifices requires more than rational debate about the scientific facts. Waste policy will remain frozen as a centralized diktat until institutions, international organizations and governments engage in a permanent dialogue on the decision-making process and develop an innovative thinking on waste.

What matters in the end is whether used resources become new resources (for nature or people) or whether they are left to accumulate as wastes (in the soil, air or water). Food waste should be rethought and reframed, as its recovery can lead to economic recovery. The waste hierarchy (prevention > preparing for reuse > recycling > recovery > disposal) should surge, shifting the economy from systematically pillaging nature to systematically enhancing it.

REFERENCES

1. http://www.time.com/time/nation/article/0,8599,1907514,00.html
2. D. Goleman, *Ecological Intelligence, How Knowing the Hidden Impacts of What We Buy Can Change Everything*, Random House Inc, 2009, p. 54.
3. A. Giddens and J. Turner, *Social Theory Today*, Stanford University Press, Stanford, California, 1988.
4. http://www.sustainabilityforhealth.org/system/documents/324/original/DB1B8533–19BB-316E-40C74D3737B2A793.pdf?1263555450
5. http://www.bionomicfuel.com/bioconversion-of-food-waste-for-energy-production/
6. http://www.neweconomics.org/sites/neweconomics.org/files/Reframing_the_Great_Food_Debate.pdf
7. http://www.worldwatch.org/nourishingtheplanet/what-works-reducing-food-waste/
8. Organisation for Economic Co-operation and Development: Working Party on National Environmental Policy, *Towards Sustainable Consumption: An Economic Conceptual Framework*. OECD, 2002.
9. http://www.container-recycling.org
10. J. Canterbury and S. Eisenfeld, *The Rise of Pay-as-you-throw*, MSW Management, 2006.

11. http://www.guardian.co.uk/environment/2008/aug/05/recycling.japan
12. Worldwatch Institute, *State of the World 2008: Innovations for a Sustainable Economy*, WWI, 2008.
13. European Commission. *Integrated Product Policy (IPP)*, available at: http://ec.europa.eu/environment/ipp/
14. E. Weizsacker *et al. Factor Four: Doubling Wealth, Halving Resource Use—The New Report to the Club of Rome*, Earthscan, 1998.
15. http://www.activedisassembly.com/

Subject Index

Note: page numbers in *italics* refer to tables, page numbers in **bold** refer to figures.